PICTURES OF TIME BENEATH

Dedication

To Bronwen, Charles, Allie and Andrew, with love, relief and gratitude.
To Jean: I wish you could have been here too.

PICTURES OF TIME BENEATH

Science, Heritage and the Uses of the Deep Past

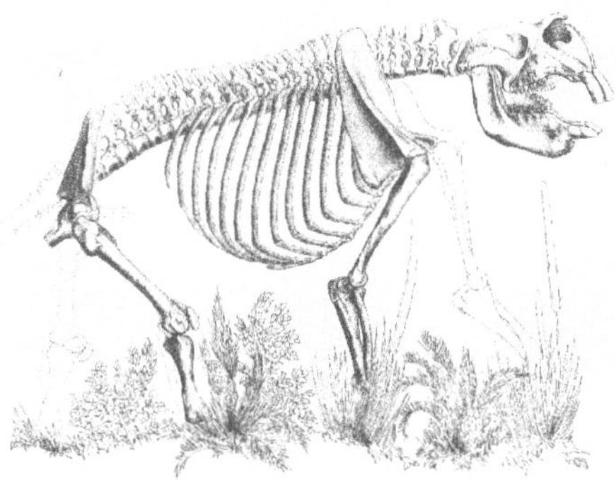

Kirsty Douglas

PUBLISHING

National Library of Australia Cataloguing-in-Publication entry

Douglas, Kirsty.

Pictures of time beneath: science, heritage and the uses of the deep past/Kirsty Douglas.

9780643097049 (pbk.)

Includes index.
Bibliography.

Cultural property – Australia.
Historical geology – Australia.
Landscape – Australia.
Callabonna, Lake, Region (S. Aust.)
Hallett Cove (S. Aust.)
Willandra Lakes Region (N.S.W.)

559.4

Published by
CSIRO PUBLISHING
36 Gardiner Road, Clayton VIC 3168
Private Bag 10, Clayton South VIC 3169
Australia

Telephone: [+613] 9545 8555
Local call: 1300 788 000 (Australia only)
Fax: +61 3 9662 7555
Email: csiropublishing@csiro.au
Web site: www.publishing.csiro.au

Cover: Diprotodon skeleton standing in grass. Richard Owen, 1877. *Researches on the Fossil Remains of the Extinct Mammals of Australia: with a Notice of the Extinct Marsupials of England*. (J. Erxleben: London). Volume 2, Plate 35. National Library of Australia NLq 569 OWE (382 7 67)

Set in 10/12 Adobe Minion and ITC Stone Sans
Edited by Anne Findlay, Editing Works Pty Ltd
Cover design by Samantha Duque
Text design by James Kelly
Typeset by Desktop Concepts Pty Ltd, Melbourne
Printed by Ingram Lightning Source

CSIRO PUBLISHING publishes and distributes scientific, technical and health science books and journals from Australia to a worldwide audience and conducts these activities autonomously from the research activities of the Commonwealth Scientific and Industrial Research Organisation (CSIRO).

Feb26_RP_ILS

Contents

Acknowledgements

Affection and thanks to Jim Bowler, who set me on this course, and to Tom Griffiths, who helped me along the way.

The first time I wrote this book it was for a doctorate in History, an expansive journey through landscapes and communities, three disciplines, 500 million years of geology and two centuries of ideas about the changing Earth. I had the luxury of exploring side paths in self-indulgent detail, letting tangents and continuities weave through my narratives to create thematic resonances rather than fix a story on a coherent theoretical framework. The process of turning one piece of writing into another has in most ways benefited from the five years of neglect that followed while I was distracted by other ideas and obligations. While less expansive and perhaps with fewer poetic resonances than its ancestor, this new book profits from extensive rewriting and un-writing, and a more mature understanding of the challenges geological heritage offers regulators, and of the historical contingencies which affect how we can recognise and protect the places people value.

For the most part I have not updated the text to reflect five years of technical improvement in geochronological tools and technologies, so some of the dates do not reflect current orthodoxy. I am more interested in relatedness than specificity, and in any case the speed of change in the disciplines would ensure that any updates I made would be out of date before the book went to press.

Pictures of Time Beneath was completed in part while visiting the History Program of the Research School of Social Sciences at the Australian National University for several months during 2009. My thanks are due to the academic and general staff and the students of the History Program for housing me, as well as for supporting the doctoral thesis which informed this current work, undertaken there between 1999 and 2003. In particular I am grateful to Pat Jalland, Melanie Nolan and Tom Griffiths for spurring me to complete the book, and Ann McGrath and Karen Smith for organising my visit.

Thanks also to my generous colleagues in the Heritage Division, who spared me for several months despite the usual heavy workload, and encouraged me to get on with it, in particular Jennifer Carter and Jane Ambrose, and to Terry Bailey and Theo Hooy for comments on a draft.

Thanks to Barry Smith, John Mulvaney, Paul Turnbull, Tom Griffiths, Bronwen Douglas and Libby Robin for feedback on drafts, and to the examiners of my thesis, Jane Carruthers, Chris Ballard and Stephen Pyne, who provided invaluable critique and encouragement. And to friends who buoyed me while writing, goaded me to meet my deadlines or shared the pain. In this capacity I am especially grateful to Donna Dening, Tiffany Shellam, Karen Riley, Ryan Walter, Ingereth Macfarlane, Emily O'Gorman and Kim Sterelny. I owe a debt of gratitude and several drinks to Brett Calcott and Martin Smith for all their help with maps and diagrams, and to John Magee, Martin Callinan, Michael Main and the National Library of Australia for use of photographs.

Thank you to John Manger, Tracey Millen, Pilar Aguilera and CSIRO Publishing for seeing virtue in publication, and to Anne Findlay for generous editorial feedback.

Thanks finally to all those who supported me in various intellectual, institutional, social, pastoral and emotional ways during and following the production of my PhD, without whom there would have been no book, or at least not this book. This includes my supervisory panel (Tom Griffiths, Paul Turnbull, Mike Smith and Richard Baker), advisers who generously read drafts like Maud McBriar, John Magee and Rod Wells, my colleagues at the ANU and in the Department of the Environment, many librarians and archivists from institutions around Australia and the world, those generous informants and collaborators whose personal, scientific and institutional memories informed and formed my stories, and most of all my friends and family, especially Bronnie, Charles, Allie and Andrew. Full references and acknowledgements can be found in Kirsty Douglas (2004), 'Pictures of time beneath': science, landscape, heritage and the uses of the deep past in Australia, 1830–2003. Unpublished PhD thesis (Australian National University: Canberra).

Introduction

SCIENCE, LANDSCAPE, HERITAGE AND THE USES OF THE DEEP PAST

The entangled country

'In making ourselves acquainted with the actual extent of our knowledge respecting the Geology, Mineralogy, Zoology, and Botany of South Australia, we are brought face to face with the fact that there are many missing pages, and even chapters, in its history.' So spoke Ralph Tate in 1878 to the rejuvenated Philosophical Society of Adelaide, borrowing Charles Darwin's narrative metaphor of two decades earlier, which characterised the history of life as a book fragmented and dissipated by time. Before the British-born naturalist Tate arrived in South Australia from London in 1875, a consciously independent indigenous science had gained a foothold in the fledgling cultural institutions and societies of the Australian colonies. Tate's presidential address of 1878 stressed the need for local knowledge and the refinement of Australian science so that it might reflect the southern continent's geological and biological singularity. Beyond the 'mere technical barrenness' of models borrowed from the Old World, the new indigenous science was to be 'real, practical, and interesting'.

In *The Custodians,* his 1997 meditation on national identity and the meaning of the past, the Australian novelist Nicholas Jose froze Tate's narrative into a panorama, with his description of geologists' conceptual genius as 'seeing through the surface of the land' to 'pictures of time beneath'. Both Tate's and Jose's conceits point to ways in which the Earth sciences in Australia have contributed to finding and restoring Darwin's missing pages, in the process building narratives which solder representations of the past to landscapes, nationalism and cultural identity. *Pictures of Time Beneath* explores ideas about the deep past in Australia since 1830, examining contemporary notions of geological heritage, cultural property, identity, antiquity and time. It brings together related themes, places and stories that knit into a narrative about the construction and interpretation of Australian landscapes as landscapes in time. From the vertebrate palaeontology collections of the Royal College of Surgeons in London to suburban Adelaide in the 1970s, from 1950s Berkeley, California, to Pleistocene Lake Mungo, *Pictures of Time Beneath* investigates the conceptual, geological, intellectual and historical continuities between emerging ideas about the shape of past worlds, the celebration of particular deep time landscapes and the nation.

A landscape is assembled from the relics of past land surfaces, by the interactions, through time and space, of geological and hydrological processes and climatic, biological, cultural, architectural and political influences. A landscape acquires meaning – aesthetic, economic and scientific – through being viewed. However, landscapes are more than the sum of their views, which are ephemeral, the product of a certain angle of the light, or of a change in attitude, for instance following the revelation that sheep thrive on saltbush after all. Any given landscape is the accumulating demonstration of change over time, and can only be approximated in the present. Past

Figure 1: Major Mitchell sketching the entrance of the caves in Wellington Valley, New South Wales. *(William Romaine Govett, 1843. National Library of Australia nla.pic–an4700786)*

landscapes can only be reconstructed through their traces left in present landscapes that are themselves built, as the American writer John McPhee suggested, of 'chaotic, concatenated shards of time'. Which is not to suggest that a landscape – in the sense of a place, at a particular time, of a particular extent – is not real and tangible, but it may be over-determined. A landscape is an expanse of countryside (in time) used as a vehicle for representations.

Until about 1830, the landscapes of British Australia did not have a deep past. Settlers regarded the flattened country with 'island-like' chains of low, rounded hills as confirming the axiom suggested by its apparently anomalous biota and its debatably primitive people, that the island-continent had emerged last from beneath the flood-waters that wiped away all but a boatload of humanity. But in 1829 George Ranken, a Bathurst settler, found the bones of giant extinct marsupials in the limestone caverns of the Wellington Valley in central New South Wales, proving that Australia's marsupial fauna, at least, had a pedigree comparable with the mammals of Europe and the Americas (see Figure 1). From then its past deepened inexorably, until the last of lands became the oldest continent, its flora and fauna adaptive instead of relict and its indigenous people ancient rather than primitive, culturally varied rather than homogeneous through space and time, providing a sterling example of cultural continuity and adaptation since the Pleistocene epoch.

During the last 200 years, stratigraphy and the sciences that rely on it – geology, palaeontology, archaeology – have changed the way people see the world, particularly in regard to the recognition and interpretation of signs of age in the earth and rocks. Their differentiation and professionalisation in the nineteenth century paralleled the backward extension of Earth history in the Western scientific canon. Curios in their own right, Australian natural history objects and indigenous material culture were in demand among European and British scientists for the correspondent glimpses they were understood to offer into the deep past of the Old World. Living Australian fauna appeared to provide proxies for European fossils of the early Tertiary period (now replaced by the Palaeogene and Neogene periods, the Tertiary spanned roughly 65 million to two million years ago, see Figure 2). Mainland Australian and Tasmanian Aboriginal people were equated respectively with the fossil Cro-Magnon and even more 'primitive' Mousterian cultures of Europe. Australian rock sequences provided important global correlations and could be 'fitted in' to European systems.[1]

Australian and New World natural history eventually provided a corrective to a predominantly Eurocentric view of creation. The discovery of fossil plants and animals in Australia's interior proved the pedigree of Australian flora and fauna. By the early 1860s, the priest and natural historian Julian Tenison Woods wrote confidently of the age and uniqueness of Australian landscapes. From the turn of the nineteenth into the twentieth century, South Australian geologists Walter Howchin and Douglas Mawson discovered the first Precambrian glacial surfaces and sediments of the Sturtian Glaciation, helping set the stage for one of the most exciting and contentious geological hypotheses of the late twentieth century. Adherents to the 'snowball Earth' theory suggest that during the late Neoproterozoic era, 850 to 600 million years ago, Earth's climate oscillated between such severe icehouse conditions that at times the entire surface of the planet might have been covered in ice, interspersed with runaway greenhouse conditions. Such extreme environments perhaps triggered the evolution and radiation of multicelled animals and later the development of hard parts like exoskeletons and vertebrae. In another discipline, physical anthropologists and Earth scientists contested human antiquity in Australia for nearly two centuries, until John Mulvaney's work at Fromm's Landing on the Murray River and at Kenniff Cave in southern Queensland during the 1950s and 1960s and the discovery of human remains at Lake Mungo in the 1960s and 1970s established an age of at least 25 000 years for Aboriginal habitation, and in the 1970s introduced the phrase 'forty thousand years' to debates about human antiquity.[2]

As colonial cultural institutions and government surveys were established from the middle of the nineteenth century, many of their directors and collectors, while still willing to send casts and duplicates of specimens to British institutions, were eager to exploit the material remains of the Australian past in the interest of developing local science and training local scientists. During this period, Australian natural sciences shifted from imported knowledge, heavily reliant on Europe and Britain for a supply of practitioners and theory, to indigenous sciences of which Ralph Tate would have been proud. The same period saw a transformation from the interpretation of Australia as a geologically young continent to recognition of the great age of some of its landscapes, landforms, cultures and biota.[3]

Landscape, the geosciences and heritage are linked historically, imaginatively and practically in the study of changing ideas about Australia's deep past, its current nature and potential. Geoscientists explain particular features as marks of age. Their interpretation allows the partial reconstruction of the shape and nature of past land surfaces, climates and ecologies, which in turn can serve an important predictive function for scientists, industry and policymakers. Age in the landscape is used for national and parochial ends, to trace continuity and discontinuity, to claim, test, contest or erase ownership, community and custodianship. Antiquity provides a surrogate for topographic grandeur or productivity.

Three decades of work on geological heritage and geoconservation in Australia have produced a considerable body of literature. This emerging corpus reveals a changing philosophy and increasing methodological sophistication. Geoconservation and geological heritage protection are still largely underpinned by utilitarian considerations, but increased public awareness

Eon	Era	Period	Epoch	Million years
Phanerozoic eon	Cenozoic era 'recent life' 65–0	Quaternary period 2.6–0	Holocene epoch	0.01–0
			Pleistocene	2.6–0.01
		Neogene 26–2.6	Pliocene	5.3–2.6
			Miocene	23–5.3
		Palaeogene 65–23	Oligocene	34–23
			Eocene	56–34
			Palaeocene	65–56
	Mesozoic era 251–65	Cretaceous		144–65
		Jurassic		200–144
		Triassic		251–200
	Palaeozoic era 545–250	Permian		299–251
		Carboniferous (Mississippian–Pennsylvanian)		359–299
		Devonian		416–359
		Silurian		444–416
		Ordovician		488–444
		Cambrian		542–488
Proterozoic 2500–545	Neoproterozoic 1000–542	Ediacaran		635–542
		Cryogenian		850–635
		Tonian		1000–850
	Mesoproterozoic 1600–1000			
	Palaeoproterozoic 2500–1600			
Archaean 4000–2500				
Hadean (informal) 4600–4000				

Figure 2: Geological table (not to scale). *(From Geoscience Australia and the International Commission on Stratigraphy)*

driven in part by geologists active in geoconservation has forced governments and regulators to consider the worth of geological sites and stories. Several state governments' registers or catalogues of sites of geological significance are available online. The Geological Society of Australia has been very active in recording and promoting geoheritage sites. Recently lobbyists have been working to include Australian places in the UNESCO-accredited Global Geoparks Network, and have sought to establish Geoparks networks nationally and in the Asia–Pacific region, using the model promoted by the European Geoparks Network. Most of these lists, registers and catalogues provide no regulatory protection, but the profile of that marketing leviathan, the World Heritage List, has fuelled domestic and international interest in particular Australian geoheritage places.

So the products of the deep past have become increasingly marketable. Landscapes regarded as wastelands are given legitimacy as sites of natural heritage by virtue of the signs of longevity they demonstrate, particularly when this involves the human past. The celebration and World Heritage classification of particular landscapes or features such as Uluru in Australia's Northern Territory as cultural landscapes, or the recognition of the Willandra Lakes Region World Heritage Area in New South Wales or the Northern Territory's Kakadu National Park for their Indigenous cultural heritage values (as well as for geological or biological significance) are part of the incorporation of human antiquity and continuity in Australia into the nation, providing the legitimacy of a deep human past. This tendency to recognise and celebrate age is not just a strange little national quirk. Our colonial and national institutions, political habits and any characteristics we might imagine we share, while having genealogical roots in other lands and other cultures, accumulated on a continent assembled in remoteness over

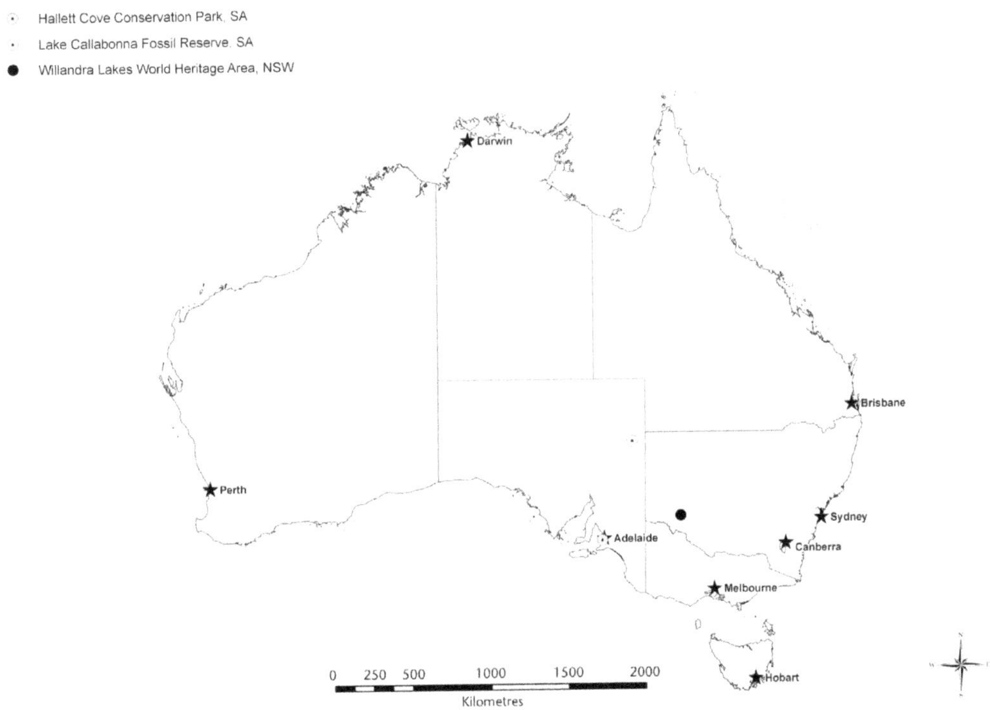

Figure 3: Location map. *(Image Martin Smith)*

tens of thousands of millennia. Australian landscapes and biota really are the products of isolation, often old and peculiar.[4]

How do we encapsulate the social–historical experiences of a nation and who chooses which histories and heritages we recognise as national? Heritage and national identity are not necessarily bound to continuity of place, even in a nation with such seemingly natural boundaries as Australia, and perhaps especially in a migrant nation. But many recent big questions, for instance about legitimacy, authenticity and sustainability, revolve around a people–land binary. Beginning with the proposition that settler nations like Australia claim legitimacy through the celebration of objects and sites which are emblematic of a pre-colonial or pre-human past, *Pictures of Time Beneath* considers how this appropriation is encouraged by the last 200 years of geological ideas in the Western scientific canon and by the nature of the continent itself, with its 50 million years plus of biogeographic isolation.

The interpretation of the material remains of the past within the academy infiltrates beyond the academy and influences the ways people think about and use the past. This permeation happens through teaching, personal communication, mainstream media, government policy and perhaps most graphically in popular science literature, film and museum displays. The iconography of this land as the oldest continent, with the oldest rocks, oldest mineral, oldest tree, life forms, multicelled animals, songbirds, rock art, funerary practice and so on may be an important part of the process of building national narratives. Particular sites and landscapes, as storehouses of information about Australia's deep past, are made scientific, regional, national and international resources. The nature of the country itself encourages us to choose or privilege certain sorts of historical narratives and heritage sites. The book focuses on three landscapes that have been celebrated as heritage landscapes and that have at various times mobilised popular sentiment towards their protection or heritage recognition: South Australia's Hallett Cove and Lake Callabonna, and the Willandra Lakes of New South Wales (see Figure 3). It tests the proposition that there is a powerful rhetorical link between collective national identity, the communication of geological ideas and the protection and celebration of material remains of the deep past by examining the ideas which first made these places scientifically relevant, the places themselves, and community or government endeavours to protect or recognise them.

Since the publication in 1831 of Ranken's Wellington Caves discoveries (shown in Figure 1), Australians and visitors to Australia have conceived, mobilised and popularised Australia's deep past, in the process memorialising particular sites, landforms, stories and objects. As vessels for representation, landscapes are historical artefacts. While landscapes can be understood to embody particular stories, they are also canvases or pages on which to paint or write regional and collective identities. Landscapes and their material productions in the fourth dimension, celebrated for the signs of antiquity encoded therein, allow nations like Australia to transcend the simple short chronology of settlement and Federation. But there are practical questions to consider: how can these landscapes-in-deep-time be managed, for whom and by whom? Who has stakes in them, and why and how are they valued? *Pictures of Time Beneath* considers these questions, the nature of geological information and its communication, its influence on understandings of Australia and being Australian, and presents a set of meditations on landscape, deep time and national identity. Any landscape is a metaphysical tangle of geological, intellectual, biological, climatic, cultural and political elements, suggesting that actual country and stories about country are built out of each other. Before examining this premise in more detail, a reflection on geology and the politics of time and place will help set the scene.

Geology, serendipity, representation and teleology

Archaeology, palaeontology and the geological disciplines depend very literally on the contingencies of data collection – the serendipitous processes of preservation (taphonomy) and discovery of fossils, features and artefacts – to construct scientific knowledge.[5] Places are codified as palaeontological sites by the presence and discovery of this data: if there are no fossils or features preserved, or if no one uncovers it and interprets it, there is no site to speak of. The logistical and financial cost of field work in remote areas is prohibitive and only the discovery, interpretation and publication of the material remains of the past justifies the institutional cost. For example, the then Berkeley graduate palaeontologist Dick Tedford expressed his delight, relief and frustration at the Lake Callabonna fossil site in South Australia, in June 1953: 'Five weeks in the field today – our 4th day on vertebrate fossils!'[6] During five weeks of field work, the expedition uncovered none of the vertebrate fossils it sought, but at the end of Tedford's first day at Callabonna, the richness and potential of this particular site was clear. Had the earlier Tertiary sites revealed significant vertebrate fossil material, the Callabonna finds would not have been as important to Tedford, and had they not found the site at all, the expedition would have been an expensive failure.

The Scottish geologist Hugh Miller was separated from Tedford by a century and a Darwinian paradigm. But in 1857 he similarly emphasised the fortuitous nature of palaeontological discovery. He explained his apparent philosophical inconsistency in shifting from a literal Mosaic understanding of creation to 'that scheme of reconciliation between the Geologic and Mosaic Records which accepts the six days of creation as vastly extended periods', in both teleological and mechanistic terms, as a consequence of the unreliability of the fossil record:

> I have yielded to evidence which I found it impossible to resist; and such in this matter has been my *inconsistency*, – an inconsistency of which the world has furnished examples in all the sciences, and will, I trust, in its onward progress, continue to furnish many more.[7]

In Earth sciences, this dependence on the fortuitous unearthing of data – contingency – controls the unfolding of geological theories. They are historical sciences, grounded in the reconstruction of present and past landscapes, past events, past relationships and past ecologies. This historical aspect, central to the presentation of geological ideas, makes it difficult to resist a narrative, if not deterministic, approach both to understanding the planet and to understanding the development of the Earth sciences.

For most Earth scientists, narrative is an unconscious representational strategy and an occupational hazard. They describe sequential relationships between – or tell stories about – the material remains of the deep past (examples follow for each case study). Borrowing from Hayden White, geology is thus both a 'discourse that narrates and a discourse that narrativises' in the sense that its practitioners adopt a self-consciously objective position 'that looks out on the world and reports it' while simultaneously they attempt 'to make the world speak itself and speak itself as a story'.[8] This narrativising tendency lends itself to appropriation by a popular media hungry for national and local myths. In contrast Henry Gee, a zoologist by training and senior editor at *Nature*, argued in the lively and iconoclastic *Deep Time: Cladistics, the Revolution in Evolution* (2000) that a cladistic approach to the fossil record – that is to say, one which rejects the search for origins, lineages and genealogical relationships between living and fossil organisms, seeking instead to approximate degrees of relatedness – is the only feasible method of mapping evolutionary relationships. He explained that 'the scale of geological time that scientists deal with' is 'so vast that it defies narrative'. Time is a 'fathomless sea' in which 'each fossil is an infinitesimal dot', whose genetic relationships are largely indiscernible.Gee argued that the fossil record is not a record as such, but shafts of light illuminating disconnected 'tableaux of unfamiliar characters' in doorways off an infinite darkened hallway 'with no landmarks to give it scale', representing deep time. The content of the spaces between the shafts of light is mostly unknowable, a useful analogy for the relationships between all species, living and fossil: 'Any story we tell against the compass of geological time which links these fossils in sequences of cause and effect – or ancestry and descent – is, therefore, only ours to make. We *invent* these stories, after the fact, to justify the history of life according to our own prejudices.' All organisms are 'cousins' in deep time, but – such is the serendipitous, incomplete and incompletable nature of taphonomy and field work – our exact relationships remain unopenable doors in the darkened hallway.

All the same, the search for links and causes continues. Perhaps, far from defying narrative, the desire of scientists to see the connections between things encourages it. Gee himself chose an analogy which implies sequence, in the idea of scenes or tableaux off 'an endless, dark corridor' (his image of dots 'lost in a fathomless sea' served him better). The search for causes, connections or origins might appear unscientific, but the advantage of geological synthesis is just that: it attempts a syncretic explanation for changes in the planet over time which is holistic as well as geographically specific, generalising from the local to the regional to the global; and it operates within an evidential paradigm. The fact that a geological scenario is speculative and not immediately subject to falsification does not necessarily render it useless or unscientific. It may well have testable consequences, which emerge in time.[9]

Since at least the late eighteenth century, Earth scientists have encouraged the production of an entirely new set of imagined spaces. Jose's description of the intellectual recovery of past environments and land surfaces by the 'X-ray eyed' geologist who sees 'through the surface of the land' to recover past worlds and lost time is part of a long tradition of pictorial or scenic representation of the geological past. The past has both depth – 'beneath' – and breadth and narrative – 'pictures'. Martin Rudwick has written widely on Earth sciences as practised in eighteenth and nineteenth-century Western Europe. In 1992's *Scenes from Deep Time,* he discussed pictorial representations of the deep past during the nineteenth century. Richard Owen and Gideon Mantell repopulated the Sussex Weald with cycads and ferns and gigantic implausible reptiles which Benjamin Waterhouse Hawkins constructed in wire and plaster for the Great Exhibition of 1851. Henry Thomas De la Beche and others sketched elaborate panoramas of ancient Dorset with a menagerie of extinct and not necessarily contemporaneous Mesozoic monsters, culminating, as Rudwick described, with Louis Figuier's *La terre avant le deluge.* Rudwick pointed out that the realism of these 'temporal panoramas' invites viewers 'to imagine that we are seeing the deep past with our own eyes, unproblematically, as from a time machine'. We can extend his example to argue that the popularity of modern museum exhibitions (and by extension, films like *Jurassic Park*) rests partly on the fact that the dinosaur-viewing public has long been educated to a cinematic appreciation of the deep past. The propensity of nineteenth-century artists to crowd their scenes with roughly contemporaneous life forms is replicated by programs like the BBC's 'Walking with dinosaurs' and 'Walking with beasts'. In De la Beche's famous Dorset panorama, giant mosasaurs prey on dolphin-like ichthyosaurs, in turn hunting extinct cephalopods, the ammonites and belemnites. Plesiosaurs haunt the shallows, resembling swan-like submarines or the Loch Ness monster. Above, pterodactyls surf the air, wary of the aquatic carnivores below.[10]

For the last two centuries, Earth scientists and museums have exploited the narrative potential of dioramas, presenting a series of linked scenes in two or three dimensions, abbreviating the innumerable variety of life to an inevitable, approximately linear progression from single-celled organisms to the pinnacle of Darwinian evolution, modern humankind. Along with Gee, anthropologists, biologists, Earth scientists, historians of science and cultural commentators have pointed out that a Eurocentric, progressive and ultimately anthropocentric view of evolution, history and the past is anomalous. Of this tendency to emphasise apparently increasing diversity and complexity as a sign of progress, the Australian geologist Ian Plimer complained in 1997 that, owing to their longevity and adaptability, 'bacteria are at the pinnacle of life on Earth, and it needs to be said that we humans have an ego incommensurate with our evolutionary place in Nature'.[11]

In contradistinction to Gee, the geologist and environmental historian George Seddon, delivering the Geological Society of Australia's 1996 Mawson Lecture on the topic of the culture of geology, pointed out that much geoscientific research 'is relational', seeking 'a comprehensive explanation that shows the linkages between superficially unrelated data'. While not peculiar to Earth sciences, this search characterises them. The relational aspect has a narrative component, in the sense that geological data often represents sequential events, but Seddon explained that even if 'sequence is of minor importance, the comprehensive explanation constitutes a "story"'. He called this diagnostic or narrative aspect 'pulling the pieces together'. Gee recognised the dangers of believing one's own conceptual short cuts and advocated an 'anti-historical' approach to palaeontology, encouraging the rejection of narrative. But for Seddon, narrative is part of the conceptual make-up of geologists and is a useful heuristic tool.[12] Add scenic representation to Seddon's relational or narrative approach and you have a visual story, or diorama, scenes from deep time knitted together into a film or slideshow. In part this difference between the two writers reduces to a difference of definition. Although Gee never defines narrative, to a certain extent he conflates it with teleology. For Seddon, narrative stands for the attempt by Earth scientists to demonstrate the relations or links between pieces of geological data. This is not to suggest that he would have rejected any less vigorously than Gee recourse to teleology or the apparent inevitability of hindsight – the conflation of sequence and posteriority with cause and design.

The literary theorist Stephen Greenblatt argued that it is a 'theoretical mistake and a practical blunder' to collapse the distinction between representation and reality, but unfortunately, 'we cannot keep them isolated from one another'. As Albert Einstein helpfully observed in 1916, organising conceits (like taxonomy, phylogeny or narrative) can 'easily assume so great an authority over us' that they become *a priori* assumptions that inform rather than reflect interpretation of the world.[13] The naturalisation of visual metaphors and narrative in the Earth sciences distances geoscientists and their audiences from the ambiguity and instability of their observations, a tendency which Gee decries in palaeontologists and which cladists aim to destabilise. Models become real; maps become country. In a nearly invisible act of semiotic slippage, stratigraphy, a fundamental heuristic tool that transformed modern geology, is both the study of the relative position of strata and the stratigraphic features themselves (the same is true of cartography). Geological maps and sections are so powerfully heuristic that their streamlining, simplification and degree of informed speculation are all but indiscernible.

Perhaps as a consequence of the cinematic, narrative or dioramic mindset, Earth scientists are prone, consciously or unconsciously, to represent geological data as 'bits of the jigsaw', as a 'rock diary', as 'many different windows in time' or as 'stories written for us in stone'. Even the term 'geological record' evokes narrative. Charles Darwin famously described such a geological record as 'a history of the world imperfectly kept, and written in a changing dialect' of which 'we possess the last volume alone' and of that, but 'here and there a short chapter' with only 'a few lines' preserved on each page.[14] Ralph Tate, an outspoken Darwinist, might well have borrowed the image that announces this introduction from Darwin's most influential monograph. *In Suspect Terrain* is book two of John McPhee's Pulitzer Prize winning quintet, *Annals of the Former World* (1998), a metaphysical and literal road trip along the Interstate 80, through a history of North American geological ideas and landscape evolution. McPhee quoted Anita Harris, a United States Geological Survey geologist, describing the geological record similarly with a literary metaphor. 'Rocks remember', she said. They are 'the record of events that took place at the time they formed'. They are books with 'a different vocabulary, a different alphabet', but they can be read. At one point Harris wanted to major in history: 'My teachers steered me into science, but I really majored in history ... I remember how amazed I was to discover, in learning how to read rocks, how much history there was.'

The calculated rhetoric of Darwin's, Tate's and even Harris' overstated literary metaphors can be traced back to the compelling imagery of Francis Bacon's seventeenth-century behest that men be equally 'studied in the book of God's word' – Scripture – 'or the book of God's works' – Nature. This is the literary version of Gee's 'dark corridor', not a rejection or refutation of narrative, but a wholehearted embracing of it. It is difficult to resist the teleology inherent in the idea of a volume, a jigsaw, an 'ever-widening window', an 'accurate picture' or set of 'windows in time' hinting at the existence of a whole, a complete picture or narrative or history, the relics, pieces or clues of which have been 'left for us' to 'piece together'. Such visual and narrative representation has a heuristic value for scientists and educators. It is a means to grasp the inconceivable.[15] This heuristic usefulness is akin to positivist historians' view of historical narrative as pieces of a jigsaw assembled to reveal the past 'as it really was'.[16] The document or archive as a repository of neutral facts about the human past compares with the notion of the geological record or fossil record as a repository of neutral facts about the planet's past.

On time

No discussion of the deep past and geological ideas can ignore the inextricability of time–space metaphors. Discussing the 'secularization of Time', which preceded the articulation of social evolutionary theory in the late eighteenth and nineteenth centuries, the anthropologist Johannes Fabian observed in 1983 two self-evident temporal 'truths in the industrializing and colonizing West': namely that

1) Time is immanent to, hence coextensive with, the world (or nature, or the universe, depending on the argument);
2) relationships between parts of the world (in the widest sense of both natural and sociocultural entities) can be understood as temporal relations.[17]

So to summarise, 'Dispersal in space reflects directly, which is not to say simply or in obvious ways, sequence in Time'.

Herbert Spencer's 'taxonomic approach to socio-cultural reality' described evolutionary relationships 'not as a *chain* of being, but as a *tree*'. Ever since, Fabian argued, 'anthropology's efforts to construct relations with its Other by means of temporal devices implied affirmation of difference as *distance*'.[18] Early nineteenth-century geological temporalising reflected these axioms as well, for instance, in the convictions of many British and European naturalists that living Australian biota represented primitive European forms. To travel to Australia was to travel backward in time, culturally and biologically. Continuing the spatial metaphor, the science of stratigraphy and the laws of superposition demanded a literal correlation of depth with age. Depth of strata 'reflects directly, which is not to say simply or in obvious ways, sequence in Time'.

John McPhee is often credited with coining the phrase *deep time*, although archaeologists, palaeontologists and geologists have long used notions of time depth as a convenient shorthand. But the term at least took popular shape in 1981 in McPhee's perceptive, sometimes brilliant *Basin and Range*, the first book of *Annals of the Former World*. McPhee likened Scottish geologist James Hutton's late eighteenth-century insight that Earth history had an unfathomably long chronology 'which, in years to come, would gradually remove the human world from a specious position in time' to Copernicus' intellectual leap in astronomy which 'removed us from a specious position in the universe'. More than a century before McPhee described the mutual entanglement of time and space with reference to geology and astronomy, the British comparative anatomist Richard Owen made a similar figurative leap, when he wrote in 1860 that palaeontology teaches 'that the globe allotted to man has revolved in its orbit through a period of time so vast, that the mind, in endeavouring to realize it, is strained by an effort like that by which it strives to conceive the space dividing the solar system from the most distant nebulae'. Like Owen, McPhee used the notion of deep time to reference the boundlessness of its astronomical counterpoint, the deep space of Carl Sagan – which is measured in years. It equally powerfully evokes the boundedness of a stratigraphic layer or an archaeological trench. To dig is to make an intellectual journey in time – which is measured in metres. Geology offers a perspective of landscape as stratigraphy waiting to happen, stratigraphy as landscape with the place all pressed out of it.

Fabian, writing of the practical advantages of a long chronology to social evolutionary theorists, similarly highlighted the inextricability of space from geological time with an allusion to astronomy: the discipline geology, he declared, 'construes Time from spatial relation and distribution' more than 'any other, astronomy excepted'. Fabian continued that anthropology's temporal discourse 'as it was formed decisively under the paradigm of evolutionism rested on a conception of Time that was not only secularized and naturalized but also thoroughly spatialized'.[19] The same could be said of the geological sciences.

Words like unfathomable, abyssal, vast and unimaginable immediately come to mind when trying to describe the age of the universe, or the planet, or even the Quaternary period (Figure 2) according to modern scientific orthodoxy. Seddon called it a 'Gee whiz' element. 'Numbers do not seem to work well with regard to deep time', wrote McPhee: 'Any number above a couple of thousand years – fifty thousand, fifty million – will with nearly equal effect awe the imagination to the point of paralysis.' In this respect Gee agreed with McPhee. In his rejection of geological narrative he described a dichotomy between lived time – '"everyday time" or "ordinary time"' as he called 'the richly interwoven tapestry of time afforded by our everyday experience' – and 'Deep Time' which is 'something far beyond everyday human experience'. But deep time could be seen as nothing other than an unimaginably great number of parcels of ordinary time stacked together. Is the difference merely quantitative?[20]

Lived time, according to Gee's explanation, is the sum of human memory and recorded history, subtracted from human experience. This ordinary time is nuanced and divisible and 'time', 'change', 'history' and 'memory' are elided in his delineation of lived experience. 'Deep Time' on the other hand is reified and amorphous. The qualitative difference Gee remarked upon stems from the 'unknowableness' of deep time, which is not made up of 'everyday experience', but of time itself. Simply stated, there are fewer markers of experience or event in Gee's deep time on which to hang a narrative. He does not address degrees of time depth: the difference between 10 000 years of deep time and 100 million. Four orders of magnitude are elided into the endless, dark corridor.

In contrast to Gee and McPhee, Seddon found that familiarity breeds, if not exactly complacency, then a certain 'matter of fact'-ness in geologists. While they 'deal every day in figures which to non-geologists may still be breathtaking', there 'can be no doubt that a constant awareness of the scale of geological time moulds the consciousness of every geologist', making it 'difficult to see our own species as an end-point, as if creation had now reached its goal and therefore come to a halt'. Notwithstanding the search for discernible patterns, an 'awareness of change as natural and continuous makes the search for constant systems seem doomed to failure'. A knowledge of Earth history and of the relative vastness of geological time compared to lived human experience 'is likely to induce a degree of scepticism about relatively simple explanations of complex phenomena', which is not to say that complex hypotheses are not regularly reduced to relatively simple models in the forums of popular science, in the interests of communication and comprehensibility.[21]

However, rather than eschew explanations or syntheses, Seddon suggested that experience validates the creation of geological 'scenarios' such as Gee derided in the sister discipline of palaeontology. As a cautionary tale illustrating the utility of supposedly unfalsifiable, non-scientific, largely speculative hypotheses, Seddon and McPhee each offered Alfred Wegener's initial thesis of continental drift of 1915. Wegener was wrong, according to both contemporaneous and later geodynamic accounts of the ancient configuration of the continents, and he proposed no mechanism for moving gigantic, fixed landmasses around the ocean, other than the image of continents 'plowing like icebreakers through solid basalt'. But his early idea provided fertile ground for the seeds of the concept of ocean-floor spreading and plate tectonics, arguably the most powerful geological synthesis of the twentieth century. McPhee quoted Kenneth Deffeyes, a lecturer in geology at Princeton, on the tendency of Earth scientists to assemble apparently unconnected phenomena into a single, synthesising hypothesis, as with plate tectonic theory: 'All science involves speculation and few sciences include as much speculation as geology.' Knowledge of the deep past may be fleeting and disconnected, but we have a lot of time to work with, and abundant, if often incongruent, glimpses, cameos or pieces of data from the long dark corridor to spin into a narrative.[22]

Fabian, still writing about anthropology but equally applicably to the geological writings of Gee, McPhee and Seddon, suggested that the 'idea of a knowledge of Time which is a superior knowledge' is an integral part of the intellectual equipment of the discipline. A related opposition can be drawn between modern secular geological paradigms and other or older ways of understanding and describing change in the landscape. This particularly seems to be the case with geologists as described by McPhee. Recasting Gee's ordinary time/deep time distinction, he wondered whether human minds have 'evolved enough to be able to comprehend deep time'. We may only be able to measure it, without comprehending what we have measured. Essentialising shamelessly and assuming a post-Enlightenment, secular, Western audience, he continued with the assertion that 'People think in five generations – two ahead, two behind – with heavy concentration on the one in the middle.' Geologists are exceptional in this respect. Trapped between the constantly shifting time scales of lived human experience and of geological time, 'they'

> wonder to what extent it is possible to absorb a set of facts and move with them, in a sensory manner, beyond the recording intellect and into the abyssal eons … They see the thin band in which are the all but indiscernible stratifications of Cro-Magnon, Moses, Leonardo, and now. Seeing a race unaware of its own instantaneousness in time, they can reel off all the species that have come and gone, with emphasis on those that have specialized themselves to death. In geologists' own lives, the least effect of time is that they think in two languages, function on two different scales.

'Cro-Magnon, Moses, Leonardo, and now' are 'all but indiscernible stratifications'. Road cuts are 'windows into the world as it was in other times'. More than a sharp division between strata, a stratigraphic boundary denotes a 'disassembled world', 40 million years of Earth history. So following McPhee, strata more than represent the passage of geological time, the mere *remnants* of former geographies. They are the essence of deep time itself.

Naturalising the geological record

There is a kind of literary evasion here. Historical and prehistorical landscapes may be landscapes of the past but they are always also landscapes *in* the present, shaped recursively by past and present understandings of time, of landscapes and of change in landscapes through time. The same applies to features of the landscape, material artefacts of the past, strata, rocks. They might be understood to *represent* particular past events in the same way as historical texts can: in the spirit of the historian Bronwen Douglas writing about ethnographic pasts, the deep past 'has gone' and exists in the present 'only as "relic"', the 'present sign of a dead thing ..., in traces of past presents more or less randomly inscribed' or preserved in the case of landscapes and geology.[23]

In McPhee's text, revealed geology constantly undermines the apparently static sociopolitical, geographical and topographical certainties of lived experience, like 'the nation-state', 'the coast' and 'the Great Salt Lake'. For geologists, and indeed for McPhee, field work represents

> the beginnings of a journey that above all else is physiographic, a journey that tends to mock the idea of a nation, of a political state, as an unnatural subdivision of the globe, as a metaphor of the human ego sketched on paper and framed in straight lines and in riparian boundaries between unalterable coasts ... a terrain crossed with geological boundaries, mammalian boundaries, amphibian boundaries.

There is a neat irony in his statement of discontinuity, change with time and arbitrariness, given the tendency of nation-states to appropriate the physical evidences of the deep past into celebratory narratives of national essence and continuity (more on that later).

McPhee's presentation of time scales is itself somewhat reductionist. While emphasising the disjuncture of scale between lived time and time 'revealed' in the stratigraphic record, his narrative does not acknowledge time outside a linear Western scientific conception of chronology and he unquestioningly accepts the broader implications of current geological orthodoxy, even where he allows space for the detail of its component hypotheses to be overturned. More particularly, like his geological interlocutors, his narrative is built on the implicit assumption that the geological time scale itself is made up of natural, consistent or logical divisions, neither constructed nor the inherited and adapted product of a particular phase of geological understanding in a particular place, namely the middle decades of the nineteenth century, when the geological record was broadly understood to show the progressive development of life forms, culminating in modern Western European civil society. For example, the Interstate 80 'displays in its road cuts rocks from every epoch and era'. It would be easy to assume from such statements that those epochs and eras were intuitively geologically determined from a scientific reality.

Like Fabian's nineteenth-century social evolutionists who took 'naturalized Time' for granted, McPhee and his geologists accepted the stages of the geological record as a presupposition of natural history.[24] The major divisions of geological time are signalled by stratigraphic disruptions like the appearance or disappearance across the globe of organisms *en masse*. Gee's approach is useful. His rejection of causality and narrative, in the interests of a more rigorously scientific approach to palaeontology, denaturalises the divisions of the geological record, while retaining them as an accessible shorthand device to show relations in time. As he explained in his introduction:

> evolutionary questions, about 'how' and 'why' – are united in implying a narrative in which causes and effects are linked: dinosaurs became extinct *because* of the after-effects of an asteroid impact, or *because* they were rendered obsolete by mammals. Chains of cause and effect may also be animated by purpose: fishes evolved legs *for* walking on land; birds evolved feathers *for* flight; human beings evolved from ape-like ancestors *because* they had bigger brains, could make tools, and use language. Popular views of science assume that cause, effect and purpose can be easily discerned.

Literary theorist Andreas Huyssen wrote in 1995 that 'the specific tripartite structure of past–present–future, in which the future is asynchronous with the past, arose at the turn of the seventeenth to the eighteenth century'. The way late twentieth-century Western cultures think about time 'is far from natural' even though it may be experienced as such. This applies equally to lived experience of shallow time (Gee's ordinary time) as outlined by McPhee and to his own and 'geologists' assumptions of the uni-directional linearity of time. Furthermore, this directional feature is easily confused with progress (both the notion of evolutionary progress which Henry Gee and Stephen Jay Gould resisted, and the notion of scientific progress), the assumption that scientific knowledge progresses either gradually by a process of steady accumulation of evidence, or catastrophically, as one false paradigm is overturned by another, more correct, synthesis or theory.

As Huyssen put it, 'the move from the past to the future has been linked to notions of perfectibility in social and human affairs that characterize the age of modernity as a whole'. He wrote of the accumulation of 'non-synchronicities' in our present, such that 'a very hybrid structure of temporality' emerges. He meant that disjunctures between the so-called developed and developing worlds, for instance, constitute different times, resonating with Fabian's social evolutionary theorists, who reasoned that dispersal in space reflects sequence in time. Huyssen's non-synchronicities can be extended, simplistically, to the ground beneath our feet. Time and change are proved in rocks and landscapes. This is a logical consequence of naturalising geological models, to the extent that a gully or road-cutting becomes a stratigraphic section, and walking 'up-section' along it can be understood as walking through time. As Tom Griffiths wrote of George Seddon, 'space and time are indivisible, dimensions of the ground at his feet, and so deep time is not an optional extra but part of the texture and poetry of place'.[25]

Geological knowledge and the junkyard of yesterday's orthodoxies

It is not always helpful to represent other narratives that explain(ed) or describe(d) the shape of landscapes and the past as 'wrong turns' or 'false leads'. Modern geological paradigms are built from a junkyard of such false leads. Trained as a geologist, I am inclined to believe in the conditional and limited truths that I was taught as a student and the corrections that have followed. But I cannot say which of these models might prove to be false leads or wrong turns or fabric for the construction of future geological paradigms. Cultural, social, intellectual and environmental histories intertwine and converge at Hallett Cove, Lake Callabonna and Lake Mungo in ways that provide elucidation of the uses and representation of the deep past through its material remains. Geological narratives coexist with other explanations of change in the landscape that do not necessarily correspond to twenty-first-century geological truths. The relationships between geological/archaeological, Indigenous, popular media and other non-secular models, and between past and present orthodoxies, are unequal ones and receive unequal publicity. Emerging from and authorised by different cultural systems, they are frequently contentious, competing and contradictory, but may resonate with or be analogous to each other in various, often unexpected and sometimes helpful ways.

Colonial naturalists and surveyors in the early and mid-nineteenth century looked for the existence of a land-locked sea or freshwater lake system in the centre of Australia, based on a combination of their scientifically informed reading of the environment, contemporary conventional wisdom about the centres of continents and wishful thinking. Eventually South Australian colonists discovered the Lake Eyre basin, which may represent (according to modern geological ideas, and to diminish a scenario of irreducible complexity to a few inadequate words) the teasing, ephemeral, salty remnant of a giant lake system, and before that, an ancient intracratonic sea.

The Kaurna people of the Adelaide Plains explained the presence of freshwater springs along the coast south of Adelaide with recourse to the tears of an old man for his nephew, Tjilbruke. That there are, or have been, freshwater springs flowing from the Mount Lofty Ranges to the coast during the Holocene epoch is not in question. The Kaurna account explains current landscape activity through the lens of the past actions of ancestors. It is causal without necessarily being teleological. Secular geomorphological models work the other way, attempting to refine past processes from their present manifestation. They tend to be synthesising and generalist rather than localising, functional and intimate.

Pastoralists and surveyors in western New South Wales and northern Victoria occasionally found fish bones and freshwater mussel shells on the oval saltbush plains interspersed among the dunefields of the mallee country. Some of them interpreted these as signs of former freshwater regimes in the now semi-arid country, speculating that the plains once were lakes. During the mid-twentieth-century and into the twenty-first, geomorphologists used these fishy remains and other evidence to construct a deep past of great climatic oscillations in these landscapes during the Pleistocene epoch (from about 2 600 000 to about 11 500 years ago), between dustbowls and well-watered lacustrine oases. The Paakantyi people of the Willandra Lakes region account for the lake landforms with an account of a hunt for a giant kangaroo, Bookamurra, who was 'tracked for days and finally caught and killed' at the southern end of the lakes. 'The lake system and surrounding area is the actual remains of Bookamurra.'[26] There is an appealing resonance in this account with the presence in the lakes of the remains of *Procoptodon*, a gigantic, extinct, flat-faced browsing kangaroo, just as nineteenth-century ethnographers and missionaries in central Australia linked the *kadimakara* of legend with accumulations of fossils in the landscape.

These narratives are all interrelated. Modern geoscientists do not have privileged access to the knowable, real, unmediated deep past. Like all of 'us', 'they' construct pasts from ill-fitting relics in the present. They have the conceptual equipment and the learned responses to identify material remains of the past with more or less precision according to modern geological orthodoxies. But these are recursive. They extend, revise or rehearse older narratives, influenced by the same data that informed other, incompatible ways of knowing about the world. It is only from within and with hindsight that they appear authoritative, revolutionary or inevitable.

To borrow from the American environmental historian Stephen Pyne, each of these places has political and cultural meaning beyond the material reality of its geomorphology, human remains and marsupial bones.[27] Australian geologists, vertebrate palaeontologists and archaeologists regard each of these areas and the suites of stories associated with them as more or less canonical.

Canonical landscapes and contested meaning

The stories of Hallett Cove, Lake Callabonna and the Willandra Lakes unfold around the revelation of a landscape's deep past in the instant of its scientific discovery, the making of a deep history akin to what Walter Benjamin, describing historical materialism, called 'that image of the past which unexpectedly appears to man singled out by history at a moment of danger'.[28] These emblematic encounters are thrown up at us from the annals of Earth sciences as the stunning, introductory, public moments that define and preface the heroic pasts of the discipline, as well as representing the beginning or emergence of knowledge about the deep past at a particular site.

Such moments and such sites are subject to contest. At Hallett Cove the details of Ralph Tate's discovery of glacial pavements in 1875 or thereabouts are cloudy, but from its public debut as a place of geological importance in 1877, the site's age proved contentious. At the time, the notion of Australian glaciation itself was deeply controversial, and many of Tate's contemporaries questioned his identification of the cove landforms as glacial, as well as his claim of their Pleistocene or Recent origin. Eighty years later urban planners and land developers challenged geologists' moral claims to the landscapes of the cove. Conservation battles fought from the 1960s to 1970s redefined South Australian public and institutional ideas of natural heritage and value in a landscape.

At Lake Callabonna the first Western scientific encounter with the lake's deep past was itself contested. Who found the bones? What does discovery mean in the context of vertebrate fossils? Who do the bones belong to? What do they mean? Longstanding questions about the marsupial nature of the giant extinct mammal *Diprotodon* could be resolved, but questions about the familiarity of Aboriginal people with the living animal and hence about human antiquity and origins in Australia remained insoluble. The lake challenges conventional notions of value, beauty and richness and its deep past reveals a potential apparently unrealised in its desolate modern landscape.

Lake Mungo's moment of danger makes it a multiply contested site. Discovered by a geologist, archaeologists interpreted and claimed the first evidence of its long human past. Aboriginal people in turn dispute academic ownership and interpretation of the material culture and human remains. Leasehold by pastoralists in the region adds a further layer of complexity to negotiations about ownership, belonging, tradition, heritage and knowledge in an already over-determined landscape, ravaged by salt, wind, water and sheep, but treasured for its wealth of pasts.

Each study consists of many stories, the intersections of lives, texts and landscape and of popular and learned, sacred and secular ways of knowing. Timing is also critical to their heritage and scientific importance. The serendipity of discovery, of a sympathetic media, museum administrator or timely government policy all contribute to public and scientific awareness and heritage significance of particular places. Landscapes, sense of place, public awareness and geological knowledge are mutually constituted.

The stories connect in the material landscapes and in the deep past. So notions of geological heritage provide a thread of continuity across the three case studies, but this means different things at different sites. At Adelaide's Hallett Cove, Earth scientists are concerned to preserve 'geology-in-action'. The landforms and the processes that shape them are all-important. The landscape itself embodies part of the continent's geological heritage. In contrast, at Lake Callabonna in South Australia's arid north-east, palaeontologists are less concerned with preserving the land surface and evidence of the processes that affect it, so much as the fossils buried within. Although the landscape is a repository of important palaeoecological information, isolation and inaccessibility limit work. It is a highly mobile landscape in any case, of shifting sands and mud and sporadic floods. Palaeoclimatic reconstruction has predominantly involved retrieving data from the sediments (bones, pollen and dirt), and removing them to museums and laboratories. The most efficient way of finding vertebrate fossils in such landscapes involves shifting huge amounts of overburden quickly, with excavators, but this presents its own problems in terms of preserving palaeoecological context.

The Willandra Lakes Region World Heritage Area falls somewhere in between the others. Archaeology is a necessarily extractive science, like vertebrate palaeontology. Its practitioners need to dig and 'salvage'. But the region contains valuable *in situ* features as well, both cultural and geomorphological, preserving evidence of landscape and climate change. Furthermore, as at the other two sites, erosion plays an equivocal role as it both destroys and reveals the material remains of the past.

Scientific, historical, intellectual and geological continuities between the three places that are the focus of this work are inescapable. The cumulative effect of people, ideas and places creates thematic and narrative resonance. It is a scientific truism that the modern shape of Australia is a reflection of Quaternary processes and underlying geology. 'Australia' during much of the Palaeogene and Neogene periods was warmer and wetter than it is today. There was little or no polar ice cap in Antarctica. As subsidence continued in central Australia and the Lake Eyre basin, uplift along the eastern seaboard and the southern margin of the basin led to renewed sedimentation. The retreat of the same Late Miocene–Pliocene sea that covered Hallett Cove's glacial pavements left relict shorelines throughout south-western New South Wales, western Victoria and south-eastern South Australia. In New South Wales, these ridges might have shaped the distribution of the Willandra Lakes. Uplift of the Mount Lofty Ranges during the Pliocene period provided a new source of sediment for the Adelaide Plains. With the onset of Northern Hemisphere glaciation during the Quaternary period, sea levels fluctuated with the growth and decline of polar ice caps, alternately covering and exposing coastal plains of the Gulf of Carpentaria, Spencer Gulf, Bass Strait, Sydney Harbour and the Great Australian Bight. At the same time, lake levels in the interior fluctuated according to the balance and distribution of precipitation and evaporation and the amount of spring run-off. Cycles of plate tectonic movement, erosion, glaciation and aridity, sedimentation and marine or lake incursion, groundwater effects, run-off and humidity, their legacies visible in landforms and stratigraphy, have combined over roughly the last 500 000 000 years to produce the living, unfinished panoramas of Hallett Cove, Lake Callabonna and Lake Mungo.

Cultural and intellectual commonalities help shape this text. The differentiation and professionalisation of the geosciences in the nineteenth century paralleled the backward extension of Earth history in the Western scientific canon. Geology, archaeology and palaeontology are historical sciences, grounded in the reconstruction of present and past landscapes, past events and past ecologies. So a history of the development of various branches of the Earth sciences in Britain, Europe and the influence of Australian material during the nineteenth century, united in the emergence of distribution studies (or biogeography), runs through the text.

Glacial studies refreshed geological and geomorphological investigation in the 1840s. The theory that much of Europe had been covered by glaciers at some time in the past rejuvenated the study of the distribution of living species and the mystery of the presence in the fossil record of giant extinct mammals, or the apparent fossil coexistence of cold-climate and tropical animals in regions where both became locally extinct. In the 1850s convincing evidence of the coexistence of primitive human beings took debate in a new direction and challenged preconceptions about human antiquity. Perhaps ancient human hunters rather than extreme cold were responsible for faunal extinction. Nineteenth-century savants like

Charles Lyell (*Geological Evidences of the Antiquity of Man,* 1863 and 1873) and James Geikie (*The Great Ice Age*, 1877) connected these three strands, the climatic, the biotic and the cultural, to speculate further on the distribution of living and fossil plants and animals. The modern debate about the extinction of many species of large vertebrate during the Pleistocene rehearses some of these same factors and helps connect the geological and intellectual histories of Hallett Cove, Lake Callabonna and Lake Mungo, in a smorgasbord of climate change, extinction and human exploitation of environments through time.

Questions about the timing and existence of an inland sea provide a common intellectual thread, both the predicted sea of western New South Wales and central Australia that never eventuated for the inland explorers of the nineteenth century, and the historic sea or inland lake system that geological orthodoxy suggests covered parts of central and southern Australia during the Miocene and earlier. The inland explorers read and misread signs of this sea in the shape of the land, in its shroud of dirt and salt, its ripple-like dunefields and its fossil productions, as well as in the advice, memories and cosmologies of the Indigenous people they regarded as informants and guides.

Geological heritage resists simple definition. Although some Earth scientists might prefer to divorce geological heritage from tourism, identity politics and nationalism, this has not proved possible. The preservation of landscapes on the basis of their geological significance alone is not enough. In Australia, protected areas and their related legislation and management regimes focus on conserving species and habitat. The states do not provide consistent legislative protection or recognition for geologically significant places in the same way that biological heritage is protected in national parks, and the contingencies of Federation, the preoccupations of the 1890s and Australia's reliance on primary industries have so far prevented the Australian Government providing clear guidelines and leadership. The Geological Society of Australia and workers in a number of state and federal geological surveys and environment departments have suggested general principles for geoheritage recognition and geoconservation management. Over more than a decade, the Australian Heritage Commission and its successors in the Department of the Environment and the Australian Heritage Council have struggled to identify sites of outstanding geological heritage; World and National Heritage listing provides legislative protection for certain places, but problems of identification, authority, administration, jurisdiction and governance remain.

As George Seddon observed, perhaps part of the answer is education and publicity. Even without a clear regulatory framework for protection, a public educated to understand and appreciate the spaces of the past is better positioned to manage them, to appreciate them and to work to protect them. Geologists' observational skills, Seddon maintained, 'assist in building a broad understanding of place', so much so that, addressing an audience of geologists, he could 'hardly conceive the image in Australia of someone who does not automatically think of the Lachlan Fold Belt when driving around, say, Yass; or when in north-west Victoria, of the Murravian Gulf and its salt beds, or in the Kimberley, of Devonian coral reefs'. Seddon valued such geological awareness 'as a basis for a more comprehensive interpretation and understanding of the present, especially in acquiring a sense of place', observing that such a geologically informed environmental awareness is fundamental 'to all sound environmental planning and important to us individually as a part of feeling at home in an environment'. So if they are to be valued by society, geological knowledge and the repositories of geological information should be celebrated and protected, 'the former through the school curriculum, the latter through safeguarding outcrops'.[29]

The Hallett Cove campaigners found it necessary to harness archaeology, social history, botany, entomology, history of science, notions of egalitarianism, government corruption and non-consultation to ensure the preservation of some of the cove's geologically important features. The success of the Willandra Lakes Region World Heritage nomination can be linked to patrimony, landscape aesthetics and cultural heritage as well as to scientific appreciation for the archaeology and geology of the region. Lake Callabonna's geological significance has long been tied to South Australian parochialism and institutional rivalry, although Ruben Stirton's mid-twentieth-century expeditions to investigate the inland lakes for their fossil potential appear to have been rather successful exercises in cross-institutional and international cooperation.

The material landscape (including fossils and features) as a repository of knowledge about the past is a national resource, inseparable from ideas about patrimony. *Pictures of Time Beneath* presents a history of the emergence in Australia of geological heritage, through an archaeology of the historically variable meaning accorded to the material vestiges of the deep past at three celebrated sites. The history of human recognition of deep-time landscapes is inescapably cultural. This contention is supported by the history of ideas, scientific literature, contemporary popular media and local histories and mythologies. Heritage is constituted through a complex interaction between the physical imprints of specific landscapes, the developing history of geological and archaeological understandings, and a parallel and intertwined micro-history of political negotiation and contested representations. The act of promoting a concept of geological heritage is itself constitutive of a new aesthetics of landscape, generating further reflection on the significance of individual sites and creating a space in which new notions of geological heritage, identity and landscape begin to be articulated.

Part One

'Like a precious cameo': Hallett Cove, geological heritage and the glaciological imagination

The saga relating the ebb and flow of fortunes on this matter is very complex

Graham Crawford, Adelaide, 1974

Treading on classic ground

Hallett Cove is no parvenu in the aristocracy of heritage. Its service to debates about deep time is part of its birthright as a longstanding site of scientific interest. From early in its scientific story, the area has been celebrated for its geological legacy. By the 1890s, Tate's Rock, a glacier-carved pavement atop Black Point (now Black Cliff), was cited as a classic site for the study of South Australian geology and the cove was a site of 'world importance' (see Figure 13). However, concerted calls for its protection from housing developments and quarrying were not made until the late 1950s. This was due to an assumption on the part of geologists and educators that the educational and heritage values of the site were inherent and self-evident and to a concurrent failure to keep abreast of altered zoning regulations rather than to any scientific neglect or devaluation of the cove, which had retained its emblematic status for South Australian Earth scientists across seven decades of controversy and investigation.[1]

Geological heritage is a many-headed beast. Its multiple possibilities emerge from the stories of Hallett Cove, read in each other and in the landscape: geological heritage as Earth history revealed through the cove itself, and geological heritage as the history of the discipline in South Australia embodied in certain of its landforms. On yet another level, the events of the 1960s and 1970s, which resulted in some of the land at the cove being reserved as a conservation park, saw the interlacing of a third layer of significance. The fight for Hallett Cove became a test case for the geological heritage movement in South Australia and its 'rescued' landforms now hold the history of that battle as well.

A National Parks and Wildlife Service (NPWS) brochure celebrates Hallett Cove as a 'precious cameo'.[2] But despite the clarity with which its landforms appear to depict their own history, like a geological theme park, the Hallett Cove landscape is not a cameo. It exists in four dimensions. Each part of the land surface at any moment can be any part of its history. Picture it in January. A sandy beach is punctuated with pebbles, stones and boulders. Low dunes on the foreshore are covered with green and yellow grass. Khaki scrub climbs to the

Figure 4: Vincennes Bay, East Antarctica. *(Photograph Martin Callinan)*

Figure 5: Hallett Cove, Holdfast Bay. *(Photograph K. Douglas)*

cliffs and salt-whipped trees cluster on the cliff top. Behind the dunes, speckled concrete foundations in the dun dirt are the legacy of beach huts, torn down in the 1970s. The ground is black, brown, mottled, striped with pink and white and red.

Now imagine its likeness during the late Carboniferous or early Permian period. Ice envelops all but the crowns of black bare mountains. A blue-white iceberg calves with a thunderous tear and splash from cliffs of ice on the edge of the sea or a glacial lake to the south-east.

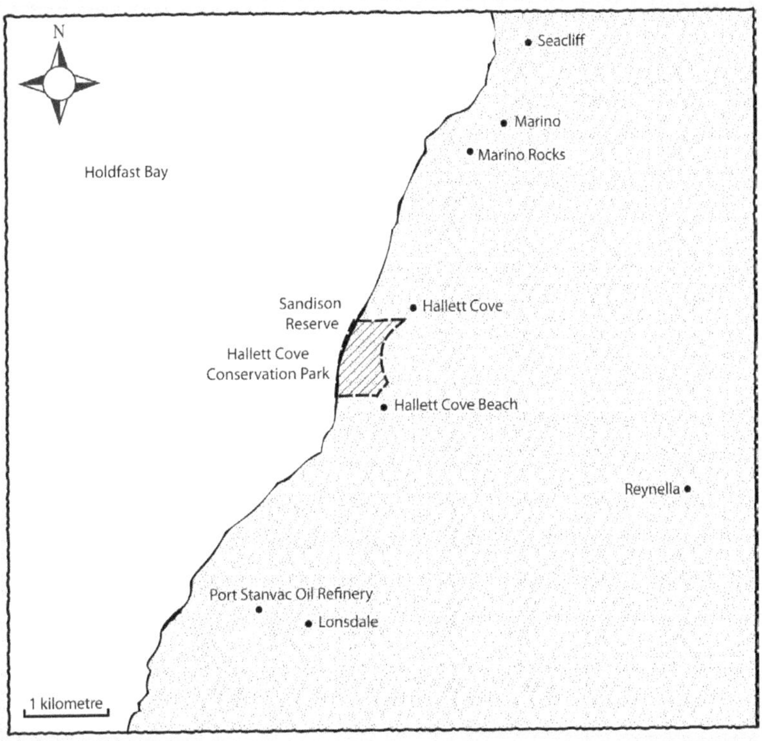

Figure 6: Hallett Cove Conservation Park and Sandison Reserve. *(Image Brett Calcott)*

As it melts, grit and boulders frozen into it fall through the water to lodge in the sediment below. 'Southern Australia' is covered by a continental ice sheet and the scene may be comparable with East Antarctica today (see Figure 4).[3]

Return to last January (Figure 5). Holdfast Bay is a soup of hopeful surfers, kelp, ecstatic poodles, assorted invertebrates, fish, children, seagulls and a ketch or two. The sun sears, but huge, scratched and polished boulders embedded in the beach sands recall the melting icebergs. Textures to taste and hear and the smell of decay verify the eyes' testimony that perhaps things in this somewhat over-determined place have changed during the upper seventeenth of Earth history. But the past is ingrained in present features. This type of juxtaposition is intellectual sustenance for scientists of the deep past who draw shallow seas from deserts, glaciers from forests and mountains from lines in the ground.

Chapter 1

PRESERVATION AND CHANGE IN AN 'OUTDOOR MUSEUM': A MODERN DIORAMA

Geological exegesis lends itself to scene-by-scene presentation. These can be visual or verbal representations, but importantly, they impose a narrative structure on the incomplete and fractured material remains of the deep past. This sleight of hand invests broken sequences of fragmented and twisted earth and rock with a determinism that simplifies and obscures both the degree of sophisticated supposition and practical science which produce geological narrative. It is a useful cognitive tool.

The present landscape at Hallett Cove is old. It has been forming and changing for more than 600 million years. But in one small section south of Waterfall Creek is represented all of the events in the following diorama summarised visually in Figure 7, and described in the following panels.[1]

Panel 1

The first panel represents the oldest rocks exposed in the park: the red mauve siltstones of the Brachina Formation at Black Cliff. These are vestiges of the formerly flat-lying sands and silts of a tidal flat on a continental shelf about 550 million years ago. Australia was then part of what is called the Gondwanan supercontinent along with the lands that now make up Antarctica, South America, Africa, Madagascar, New Zealand and India. For millions of years previously sediments eroded from the low land to the west into a series of shallow seas – the large sedimentary basin now known as the Adelaide Geosyncline – which might have extended from the present position of Kangaroo Island, beyond Lake Eyre, into Central Australia. In the process of filling the depressions, these sediments subsided under their own weight. As their water content reduced, sands in shallower water became the sandstones and in deeper water silts and muds became the siltstones and shales of the tidal flat. Eventually over the next 100 million years, the deeply buried sediments were altered by heat and pressure into the hard rock exposed at the cliff line today. Life had not yet colonised the rocky land, but algae and multi-celled organisms like jellyfish appeared in the sea.

Panel 2

The Black Cliff sediments are also the modern expression of my second panel (see Figure 8). They are the eroded remnant of a colossal mountain range, similar to the modern Himalayas, which formed in the Ordovician period, some 480 million years ago. The land that would become South Australia was north of the equator. Large-scale crustal movements buckled the sediments of the Adelaide Geosyncline (Panel 1) in a north-westerly direction. They were folded and uplifted into a massive chain sometimes referred to as the Delamerian Mountains after the Delamerian Orogeny that formed them (Figure 8). This chain stretched south into

Figure 7: Geological section, Hallett Cove Conservation Park. Solid white lines represent major unconformable boundaries, dashed lines in the lower half of the image outline folds of the Delamerian Orogeny. *(Photograph K. Douglas)*

Figure 8: Shore platforms at Black Cliff showing folding. *(Photograph K. Douglas)*

Antarctica and north into central Australia. Life continued to diversify during the preceding Cambrian period and, although there was no life on land yet, the oceans contained trilobites, jellyfish, molluscs and sponges.

Panel 3

My third panel – a diorama in itself – is represented at Hallett Cove by an absence. While grounded very literally in the evidence provided in the earth and rocks, geological exegesis also relies on concepts as esoteric as unconformities. An unconformity functions rather like parentheses. It is a contact between sediments of different ages that delimits a period of non-deposition or erosion. It is Charles Darwin's 'history of the world imperfectly kept'. This particular unconformity defines the base of a palaeovalley and represents more than 200 million years of landscape evolution. The Delamerian Mountains were intensely eroded, leaving no hint of these changes as they affected the cove. But elsewhere in Gondwana, central Australia was covered by shallow sea. Fossils of bivalves, crinoids, graptolites and the first jawless fish appear in mid to late Ordovician sediments. Vascular plants colonised the land. There was volcanic activity on the 'east coast', as eastern Australia formed during the Silurian. The first jawed fish appeared and diversified in Devonian seas, as did tree-sized club mosses on land. Forests appeared in swampy areas. Early amphibians and insects inhabited the land. Australia moved south of the equator. The Carboniferous began warmly, but cooled as Gondwana moved across the South Pole. Conifers evolved and insects took to the air. Glaciers began to develop.

Panel 4

The fourth panel is once more represented in the rocks at Hallett Cove in the wide band of glacial sediments which fill the palaeovalley from Panel 3. By 280 million years ago, change in the configuration of the Gondwanan lands and the sea had influenced global weather patterns. Lower temperatures and expansion of the polar ice caps ensued. Gondwana was gradually covered by a wet-based continental ice sheet, which in Australia blanketed much of the south-east. This is now known as the Permian, or Gondwanan, glaciation. The flattened view might have looked something like the Antarctic ice sheet today, had there been anyone to see it (Figure 4). As it flowed, the glacier tore rock and sediment from the land surface beneath. The resulting debris, frozen into its base, gouged, scratched and polished the denuded bedrock (see Figure 9; Figure 13).

Ten million years later, during the middle Permian, the climate was warming and the ice thinned. Melt-water flowed from the surface and from under the ice depositing fine rock flour and coarser material. This accumulated in layers on the floors of melt-water lakes, as at Hallett Cove, as the glacier moved westward and north-westward. Floating icebergs, calved from the ice sheet, shed debris and rocks, or 'dropstones', into the sediments of the lake floor (see Figure 9).

The vocabulary of glaciers affirms John McPhee's declaration that geology is indeed a fountain of imagery.[2] Hallett Cove is no exception. Since the Permian period it has been a premier site for dropstones, erratics, lunate fractures, crescentic gouges, chatter marks, striated bedrock, roches moutonnées (named for their resemblance to recumbent sheep), lodgement till, flow till, rock flour and tenacious clay. Verbal deposits aside, the presence of these features

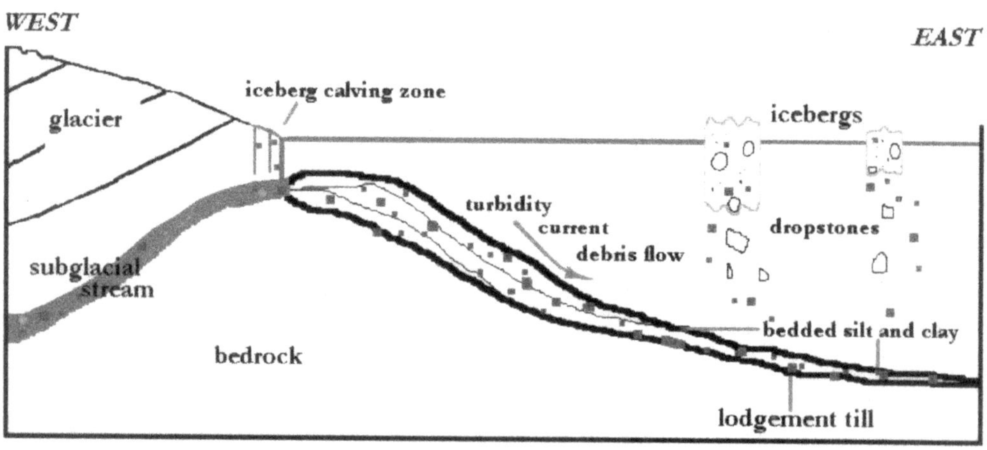

Figure 9: The formation of striated pavements, glacial tills and erratics in a glacial lake. As the glacier moved west, the bedrock beneath was abraded, leaving striated pavements like Tate's Rock. Icebergs calving into a lake at the retreating end dropped boulders and gravel as they melted, which today form part of the beach at Hallett Cove and are part of the erosion from the cliffs. *(Adapted from Giesecke, 1999)*

Figure 10: Erratic boulders eroding from cliffs above Hallett Cove. *(Photograph K. Douglas)*

at the cove provides evidence of the passage of a glacier and its direction. The Proterozoic rocks of Panel 1 were grooved and gouged by the ice sheet and friction marks – lunate fractures, crescentic gouges, chatter marks, striations – remained after the ice retreated. Larger rocks trapped in the glacier or in icebergs could be transported a long way, before being dumped when the ice melted. These include the glacial erratics that rest on the beach at Hallett Cove, that erode from the glacigene cliff sediments to the south, and rest on the cliff tops above (see Figure 10, Figure 12).

Panel 5

Above this, another unconformity marks another filled palaeovalley, a contact between the glacigene sediments and Pliocene sediments above. This contact represents another 200 million years of erosion or non-deposition. As Hallett Cove eroded during the Triassic, dinosaurs appeared. Inland Australia was covered by seas inhabited by fish and plesiosaurs. Conifers, cycads and fern prairies were common on the land. Australia and Antarctica started to separate during the Jurassic. New coastlines formed as the rifting which had begun to split Gondwana apart at the end of the Permian separated Australia and Antarctica completely by about 45 million years ago during the middle Eocene. Flowering plants, mammals and birds diversified after the decline of the dinosaurs. Many modern groups of Australian biota had appeared by the Oligocene. At the close of the Miocene, about five million years ago, the climate was warm and wet.

Panel 6

My sixth panel is set four million years ago, in the Pliocene. A warm shallow sea briefly covered low-lying land to the east of the modern Gulf St Vincent. Sediments and seashells accumulated on the sea floor, together with shellfish and the many-chambered, calcium carbonate 'shells' of single-celled marine organisms called foraminifera. These hardened to a fossiliferous sandstone, exposed two million years later, when the present Mount Lofty Ranges were uplifted. The sandstone is exposed in the Amphitheatre and also forms a bluff along the north-western part of the park and is the modern embodiment of this panel. The next two million years or so are again not represented by the Hallett Cove rocks. Australia became warmer and drier as its northward drift continued. Abundant rainforests were replaced by grasslands and eucalypt woodland. Grazing mammals diversified.

Panel 7

In the Pleistocene epoch, dry periods in Australia were interspersed with wet ones. During the latter, high ground eroded. As the Mount Lofty Ranges rose, sand, silt and clay washed down to the coast in streams. These thick alluvial sediments were deposited above the exposed fossiliferous sandstone. Where the land had remained above water during the Pliocene, they blanketed glacigene Permian sediments. The unconformity between is represented by Panel 5. Seasonal changes in precipitation leached calcium carbonate from the soil, which was redeposited near the ground surface in a hard layer of calcrete.

Sea level fluctuated as the Pleistocene glaciations of Europe and North America swelled and declined according to oscillations of the polar ice caps. Glacial advances correlated with sea level retreat and interglacials with marine incursions over the continental shelf. Only small areas of greater Australia – parts of Tasmania, the highlands of New Guinea and the Kosciuszko Plateau – were glaciated even at the height of the Pleistocene ice age. In fact, Pleistocene Australia could instead be said to have experienced 'dirt ages'.

Reduced precipitation and spring run-off from the highlands made a dust bowl of the interior. Plains of 'drift' – formerly attributed to great floods or glaciers[3] – blew across the continent, forming characteristic dune systems.[4] Sea level changes sculpted the coastlines. Upstream of their drowned deltas, rivers became coastal estuaries as the sea invaded. Coastal plain became shallow sea and plain again. The inland deltas of the Riverine Plains and the Lake Eyre basin dried out. Giant marsupials browsed the forests, grazed the plains and died horribly of thirst, trapped in the mud of drying lakes. Towards the end of the Pleistocene, the ancestors of modern Aboriginal people witnessed and adapted to these changes.

Panel 8

With the retreat of the ice caps around 11 500 years ago, the sea again began to encroach on the coastal plain of Gulf St Vincent. During the Holocene it rose more than 130 metres. Kangaroo Island and Tasmania became islands once more. The Pacific Ocean flooded the Parramatta River delta to form Sydney Harbour. Waves cut cliffs into the stumps of the Delamerian Ranges, left a rocky platform at their base and removed quantities of soft glacial sediment from the beach, to expose Permian erratics at Hallett Cove. Uplift of the land surface enhanced erosion of Permian and Pliocene sediments by the creeks which washed down from the Mount Lofty Ranges and carved the badlands features of the Hallett Cove Amphitheatre. Low dunes formed on the beach. People of the Adelaide Plains occupied sunny sites along the cliffs, in the lee of the prevailing winds. These cycles of erosion, deposition and uplift continue today (see Figure 11).

The Hallett Cove Conservation Park comprises 50.48 hectares on the eastern side of Gulf St Vincent, about 20 kilometres south of the city of Adelaide, in the Hundred of Noarlunga. It is partly bounded to the south by the 3.93 hectare Sandison Reserve and partly by the low-water mark along the beach at Hallett Cove. The remainder of the park and the reserve are bordered

Figure 11: The Sugarloaf and Amphitheatre, Hallett Cove Conservation Park. *(Photograph K. Douglas)*

by residential development which makes the site unusual in South Australia's extensive system of metropolitan conservation parks. No visible boundaries separate the park from the Sandison Reserve, but a plaque marks the spot where the reserve was dedicated to the National Trust of South Australia in 1960. The trust still owns the reserve and the park is managed by the South Australian National Parks and Wildlife Service. The reserve was opened to the public in 1965 and the present boundaries of the park were dedicated a decade later. The park consists of low hills above variably rugged coastal cliffs, a grassy foreshore area near the southern boundary, an extensive rocky shore platform and sandy cove area, with deeply eroded gullies, together with westward-trending watercourses. Much of the native vegetation was cleared during the late nineteenth and early twentieth centuries and is retained in dwarfed form only in pockets of the less-accessible steep areas and in the gullies, although extensive replanting of locally indigenous flora has proceeded in recent years. Large boulders, often as big as cars, lie on the beach (see Figure 12).

The area was acquired primarily to protect vulnerable geological, archaeological, botanical and historical features that would otherwise have been destroyed during urban expansion in the 1960s. These include the 'internationally recognised' glacial pavements and erratic boulders identified in the 1870s and 'unique' geology which bears 'witness to six geological periods, each differing in climate due to Australia's changing position on the Earth's surface'. [5] From the time of Hallett Cove's scientific discovery in 1877, this refrain of uniqueness, international renown and significance has been regularly replayed by naturalists, scientists, journalists, residents and activists. Almost a century later an advertisement on behalf of the Save Hallett Cove Committee in September 1971 exhorted readers to 'Save Hallett Cove' as the 'World famous geological treasures of this unique cove are threatened!' Journalist Stuart Cockburn, a strong advocate for preservation of the cove's geological wealth and a reporter with the main Adelaide newspaper, the *Advertiser*, explained that 'Hallett Cove has, of course, long been known throughout Australia and overseas as a geological treasure house.'[6]

A field guide to Hallett Cove, published by the South Australian Museum in 1970, reiterated that the 'unique locality, acknowledged by experts to be the best of its kind, is visited by scientists from all over the world'. The South Australian Science Teachers' Association (SASTA) wrote in 1971 that 'Hallett Cove is unique ... Because of its Permian glacial deposits the region has worldwide significance', restating later that 'It is unique. It is of world renown.' The South Australian Town and Country Planning Association described the cove as 'one of South

Australia's most valuable resources of this type – a unique landscape unveiled by erosion to give almost unbelievable glimpses of the past'. The same authors explained in the *Sunday Mail* that 'overseas geologists visit the area because of its uniqueness and international importance as a link in the theory of continental drift' and the 'glacial areas are quoted in many overseas texts'. The Friends of Hallett Cove, along with members of the South Australian Museum and the University of Adelaide Geology Department provided guided tours during the 1960s and 1970s, provoking such superlatives as 'unique features' and 'world famous' from reporters.[7]

An updated field guide to Hallett Cove, published in 1999 by the Field Geology Club of South Australia in the spirit of the museum's 1970 edition, claimed that 'The Cove is recognised world-wide for being one of the most important and visibly interesting records' of the Permian ice age. In 2001 the National Trust of South Australia's journal, *Heritage Living*, described the area as 'a Mecca for geologists and naturalists since 1877'. But Ralph Tate's last word on the matter at the fifth meeting for the Australasian Association for the Advancement of Science, held more than a century ago in Adelaide in 1893, still encapsulates the celebratory nature of South Australian geologists' reactions to Hallett Cove: 'Hallett's Cove remains unique in respect of the magnitude and completeness of the glacial features which are there preserved ... we may congratulate ourselves.'[8]

Hallett Cove's colourful municipal history has left a residue of names, remembered events and heritage sites that invest the landscapes with local meaning beyond the essentially scientific and aesthetic significance of the conservation park. The cove itself was named for John Hallett, a prominent South Australian pastoralist and early Parliamentarian, who reportedly first saw the feature in 1837 while searching for missing stock. Hallett arrived in the new colony on the *Africaine* in 1836, aged 32. Land Grant records show the first sale of land to him on 14 March 1840. Alfred Hallett, his brother, joined him in 1841. John Hallett gave evidence at a court held at Kingscote, Kangaroo Island, in the late 1830s, that the cove witnessed smuggling as, to escape excise, traders put spirits and other goods ashore there and at Brighton at night by boat from Kangaroo Island, then took them north to Adelaide by bullock dray.[9]

Prospecting was carried out in the 1840s on behalf of investors anxious to partake of the colony's new mineral wealth. In 1847, prospectors discovered copper ore at Worthing Farm, a property owned by the Hallett brothers, in the valley of the Field River, 800 metres south-east of Hallett Cove proper. Samples sent to London were deemed promising and the Worthing Mining Company was incorporated on 2 July 1849. The venture remained something of a family affair: ten men, including two Halletts, each contributed £1000 to purchase the land from Alfred and John, who retained a considerable vendor interest. Alfred managed the mine. The company appointed a colonial committee, which included John Hallett, and the directors in England included several other Halletts. Mining operations commenced in 1849 when Captain John Richards and a team of five working miners arrived from Cornwall to sink the first shaft. A steam engine, boiler and other machinery arrived from England late in the year and the team erected the engine early in 1850.[10]

During the 1850s, a community of about 100 people rapidly became dependent on the mine. The village comprised a blacksmith's shop, ten miners' cottages, a powder magazine, storehouse and offices and a large manager's house, inhabited for a while by Alfred Hallett and his family. By 1855 the settlement supported a schoolmistress. The company sank three shafts half a mile apart, but the mine was never very successful. Contributing factors included low-grade ore and the hardness of the ground which made digging shafts more difficult than anticipated. The mine's proximity to the Field River and the sea made it vulnerable to underground water, which poured in from fissures at 156 feet (48 metres), the depth of the main shaft. These agents were abetted by the exodus of miners to the lucrative Victorian goldfields in 1852 which precipitated a chronic labour shortage and inflated wages for several years. Notwithstanding its importance to the small community, the Worthing Mine closed in 1857. By 1861, the cottages were described as dilapidated and continued in increasingly derelict state until demolished over a century later. Some of the miners purchased land from the company and set up small farms. Alfred Hallett died in 1877, the year Ralph Tate announced the discovery of glacial features at Hallett Cove. The first subdivision inland of Hallett Cove was laid out in June 1913 and known as Hallett's Cove Estate. In August 1914 another subdivision was laid out at Hallett's Cove proper and the Hallett's Cove Model Estate was laid out in November of that year.[11] The 'Worthing or Hallett Mine, Enginehouse and Chimney and Environs' were nominated to the South Australian Register of State Heritage Items in September 1978 and registered on 24 July 1980, a remnant of South Australia's industrial and geological history. The site is National Trust classified. Although the mine never contributed significantly to South Australia's economy, its structures include the oldest surviving engine house and associated stack in Australia. It is a relic of the earliest period of base-metal mining in Australia and the most intact disused mine in metropolitan Adelaide.[12]

These events, trivial or momentous depending on individual or collective experience, actively shape meaning at Hallett Cove as surely as the abrasion of glaciers, the swelling of mountains, the tearing of continents and the curiosity and ambition of scientific men. And the success of the 'Save Hallett Cove' campaign of the 1960s and 1970s was as dependent upon the actions, experiences and dedication of communities, garage operators, mothers of four, schoolchildren and the scientifically illiterate as it was upon any meaning or value inherent in the rocks and sediments of cove landscapes as deciphered by the geological community. Experience shapes interpretation.

This is deep time and ordinary time at Hallett Cove in broad brushstrokes. But it is not just the story in the sediments or its commonplace colonial history that makes the site iconic to South Australian geologists. Hallett Cove is what the Geological Society of Australia terms a classical site – a canonical location associated with the development and articulation of geological concepts – in this case, nineteenth-century glacial studies in Australia.

Chapter 2

'BENT UPON COVERING THE WHOLE CONTINENT WITH ICE': GLACIAL STUDIES IN THE NINETEENTH CENTURY

What a grand new feature all this ice work is in Geology? – How old Hutton would have stared![1]

Natural science and the invention of the Ice Age

Much nineteenth-century interest in the advance and retreat of glaciers stemmed from attempts to explain the same phenomena that concerned naturalists engaged in debate about the nature of other changes in the geological record. These phenomena included unusual associations, extinction in the fossil record, 'diluvium', erratic boulders, unexplained scratches and grooves on rocks. In *Theory of the Earth* (1795), the Scottish geologist James Hutton described the motion of former glaciers as accounting for boulders in alpine regions of Scotland. Others, through independent observation, increasingly reached similar conclusions.

Then in 1834, Jean Louis Rodolphe Agassiz, the 27-year-old Swiss-born naturalist, former student of the great French comparative anatomist Georges Cuvier and recent adherent to Charles Lyell's ice-rafting theory, heard of the glacial theories of his former teacher, the Swiss naturalist Jean de Charpentier (also known as Johann von Charpentier) and a civil engineer called Ignace Venetz. In the summer of 1836, a sceptic, he travelled to Bex in the Rhone Valley. Once there, confronted by the landscape and Charpentier's explanations in the field, he changed his mind. On 24 July 1837, as President of the Swiss Society of Natural Science, Louis Agassiz presented a paper – later published as the *Discours de Neuchâtel* – which posited widespread glaciation in the recent past. By proposing cyclical cataclysmic cold his glacial theory accounted for unexpected associations in the fossil record and for the phenomenon of extinction. The former great extension of glaciers explained the presence in the landscape of many features previously difficult to rationalise. He compared the planet to an organism – initial vigour and high temperature, followed by a stepwise decrease in temperature and vigour through several ages, each of which was separated by a period of stasis, when 'Death enveloped all nature in a shroud'.[2]

It was the ultimate anti-transformist, anti-gradualist, catastrophist manifesto, with cyclical, cataclysmic glacial periods in place of cyclical, cataclysmic deluge. He did not admit of any kind of evolutionary model, but insisted on the immutability of species and complete faunal replacement between each successive age. He saw the unfolding of life as progressing towards humankind in discontinuous steps and on this he appears not to have wavered during his career. In 1837 he met with certain opposition. The meeting reportedly turned into a heckling match between the great German naturalist Leopold von Buch and Agassiz, who soldiered on nonetheless. Despite opposition to the detail of the theory, Agassiz's numerous field trips, papers and lectures established glacial studies as a credible branch of the natural sciences.

On one of these field trips in 1838, British geologist William Buckland accompanied Agassiz to the Alps but remained unconvinced by his friend's 'snowy humour'.[3] Mainstream acceptance in Europe was perhaps catalysed in 1840 when Agassiz published *Etudes sur les glaciers*. Here he added the proviso that human beings did not appear until after the last catastrophe. Also in 1840, the previously equivocal Buckland publicly accepted the efficacy of glacial theory for explaining changes on Earth, including phenomena hitherto explained by a widespread or universal deluge.[4] However, despite Buckland's public capitulation, acceptance in Great Britain came slowly. As James Geikie wrote to Darwin in 1876,

> In speaking of the glacier theory having preceded that of icebergs I was referring to the fact that Agassiz was the first to point out that glaciers had formerly existed in our islands. Up to that time (1840) I do not think anyone had suspected that this was the case. But after that time the iceberg theory came prominently forward. No doubt it was not for a number of years subsequently that what is now called the glacier theory came into vogue.[5]

Even Charles Lyell accepted that glaciation had once been more extensive, but he baulked at Agassiz's non-gradualist approach to faunal replacement and geological change. To Lyell, it was not a simple case of replacement of Great Flood with Great Cold as an 'ideal force'. He was unwilling to compromise his uniformitarian paradigm. Until conclusive evidence of the great extent and destructive power of modern, continental-scale ice sheets at the poles – as a contemporary geological force operating in the modern world and hence assimilable into Lyell's doctrine – was published, he struggled to come to terms with the notion. In 1869, writing to Alfred Russel Wallace, he argued that

> until we obtain a more certain knowledge of the manner in which ice can erode rocks & the rate at which it can cut through & above all, its power to throw out the broken rock or sand or mud to which it gives rise by erosion we shall be in the dangerous possession of an ideal force which like the old fashioned 'convulsion of nature' or catastrophe or diluvial wave of our predecessors is able to effect whatever it is required to do & to supercede almost every other agency.

However, once convinced, he embraced it as further proof of the operation of gradual forces to effect changes in the planet.[6]

Agassiz remained a catastrophist. In 1866 (*Geological Sketches*) he extended his theory to glaciers of the New World, gleefully using such evidence to strike a blow against transformists like Darwin, by demonstrating that cold had eliminated life in the fairly recent past all over the globe. Darwin told Lyell in September of 1866 that 'I quite follow you in thinking Agassiz Glacier-mad' and lamented that a man of his energy and 'splendid imagination' had not been raised a gradualist: 'What capital work he would have done if he had sucked in your *Principles* with his mother's milk – It is wonderful that he should have written such wild nonsense.'[7] The Harvard botanist Asa Gray wrote to Darwin in August of the same year that Agassiz was 'bent upon covering the whole continent [of America] with ice' to refute the continuity of life 'from tertiary or post tertiary period to ours', and with it gradualist and transformist theories of change in the fossil record, particularly Darwin's theory.[8]

Darwin himself supposed at the time that 'the whole world was cooler during the Glacial period' but, based on vegetation of New Zealand and the west coast of South America, 'where Glaciers now descend to, or very near to the sea', felt it 'rash' thereby to conclude 'that all tropical forms would be destroyed by a considerably cooler period under the Equator'.[9] He also felt ambivalent about possible mechanisms for survival of tropical species in a fully glaciated

Figure 12: Erratic boulders on the beach at Hallett Cove. *(Photograph K. Douglas)*

globe, and debated with and about the English botanist Joseph Hooker on the same topic, as for example, in a letter to Lyell of 16 February 1866. Hooker insisted that 'certain orders of plants' which had survived to the present could not have withstood the cooler climate, a difficulty Darwin acknowledged with playful reluctance:

> I do not at all suppose that nearly all tropical forms were exterminated during the cool period ... Hooker's paper in Nat: Hist: Review is well worth studying, but I cannot remember that he gives good grounds for his conviction that certain orders of plants could not withstand a rather cooler climate ... We have only just learnt under how cool a temperature several tropical orchids can flourish. I clearly saw Hooker's difficulty about the preservation of tropical forms during the cool period, & tried my best to retain one spot after another as a hot house for their preservation: but it would not hold good.[10]

So the reality of widespread past glaciation, at least in the Northern Hemisphere, was broadly accepted among the European geological fraternity by the late 1850s, though how extensive, and exactly what its effects had been was debatable. For much of the rest of the century, the topic exercised the intellectual imagination of geologists, botanists, comparative anatomists, zoologists and physicists alike: 'Interesting as many chapters are in the history of the world I do not think that any one comes nearly to the glacial period or periods', wrote Darwin to Geikie in 1876. Speculation, public and private, about the causes of glacier extension, was rife.

Resolving 'the agencies in operation'

The difficulty remained of finding a mechanism. James Croll, an amateur philosopher and scientist, working as a caretaker at the Andersonian College and Museum in Glasgow, published his theories on Earth movement in 1864 when he was 43.[11] Later, from 1867 onwards, he was the keeper of maps and correspondence in the Geological Survey of Scotland. Croll examined ocean currents and temperatures, trade winds and minute changes in sea level to come up with a fairly integrated account of glacial advance and retreat through time. He proposed that the planet has cycles of tens of thousands of years wherein it varies minutely in its course over time due to three factors:

- eccentricity of its orbit;
- precession of the equinoxes (change in the direction of Earth's axis as it turns around the axis of the ecliptic);
- axial tilt.

Such phenomena cause fluctuations in the amount of solar radiation reaching the planet's surface. Reduced radiation corresponds to colder temperatures and glacial advances. Increased radiation corresponds to warmer temperatures and glacial retreat. 'Agencies brought into operation' by the increased eccentricity of Earth's orbit (its variation from almost circular to elliptical) result in changes in the strength and direction of trade winds, which in turn influence ocean currents and the differential heating and cooling of the hemispheres.[12] In effect, Croll proposed long secular summers and winters, 'depending as truly as the annual ones do upon planetary motion and, like them, also fulfilling some important end in the economy of nature'. Annual seasonality would compound the effects of secular seasonality. When the 'excentricity is about its superior limit', physical causes combine to '*lower* to a very considerable extent the temperature of the hemisphere whose winters occur in aphelion, and to *raise* to nearly as great an extent the temperature of the opposite hemisphere, whose winters of course occur in perihelion'.[13] Perihelion is the point in its elliptical orbit that Earth is nearest to the sun. At aphelion, it is furthest away.

In 1868 Croll reflected in a letter to Darwin on the 'curious circumstance' that 'all the agencies in operation should tend to one effect viz the production of snow on the one hemisphere and the melting of it on the other'. He called it 'the grand secret of the Glacial Epoch', whereby the 'heat was all pushed to the one side of the equator and the *cold* and *moisture* to the other side'. Even 'the very nearness of the sun during summer tended to increase the amount of snow falling'. By extension, contemporaneous Southern and Northern Hemisphere glaciation is impossible. When the Northern Hemisphere is icy, the Southern enjoys a gentle summer of many generations' duration and 'the entire mass of ice which presently exists in the Southern Hemisphere would be transferred to the Northern, leaving the quantity of liquid water unchanged'.[14]

Croll's theory was widely discussed in the 1860s and 1870s. Sir William Thomson, Lord Kelvin, accepted many of Croll's ideas, which had implications for arguments about the age of Earth. Lyell was not entirely enthusiastic about secular summers and winters and the notion of a warm Southern Hemisphere while the Northern was glacial. In 1869 he worried to Wallace that Darwin had not taken enough account of his objections, or thought Croll's theories through. But from 1868, Darwin produced an extensive correspondence on the topic, with Croll himself, with Lyell, his botanist friend Joseph Hooker and with James Geikie, among others. As he wrote to Hooker in November 1868, having 'read Croll', a 'wonderful man', he was 'almost' convinced that the level parts of North America 'must have been covered with [a] sheet of glacier ice'. He wrote to Croll the same month 'about the alternation of glacial & *warmer* periods in N. & S', hoping, as he told Hooker, to remove the botanist's 'objections to cool period extending to Equator', by allowing 'temporary migration of equatorial flora northward'. He had thought himself 'compelled' to assume the '*whole* world was simultaneously cooler. It will be an immense relief.'[15]

Although running contrary to evidence Darwin had observed in the field, Croll's notion of secular seasonality helped the naturalist solve the difficult issue of geographical distribution of plants and animals and their re-population after a glacial period. In 1868 he asked Croll:

> Am I right in supposing that you believe that the glacial periods have always occurred alternately in the Northern and Southern Hemispheres, so that the erratic deposits which I have described in the S. parts of America, and the glacial work in New Zealand could not have been simultaneous with our glacial period. From the glacial deposits occurring all round the Northern Hemisphere, and from such deposits *appearing* in S. America to be as recent as the north, and lastly, from there being some evidence of the former lower descent of glaciers all along the Cordilleras, I inferred that the whole world was at this period cooler. It did not appear to me justifiable without distinct evidence to suppose that the N and S glacial deposits belonged to distant epochs, though it would have been an immense relief to my mind if I could have assumed that this had been the case.[16]

Broadly speaking, Croll's astronomical ideas accord with modern scientific orthodoxy about the advance and retreat of ice within the longer cycles of major glaciations, but in reverse. Although widely discussed in scientific circles in the 1870s, they fell out of favour in the later nineteenth century when weight of evidence began to suggest that the last ice age ended around 10 000 years ago, not 80 000 as Croll's model suggested. Between 80 000 and 100 000 years ago, when according to Croll's model the planet should have been warming, it was again plunging into an ice age.[17] Despite its flaw, the astronomical model provided an important beginning to understanding cycles of glaciation. More of the answer was to be found in ideas about glacial climatic equability. Croll assumed that extreme temperatures trigger ice ages: the hemisphere in which winter occurred at aphelion enters an ice age, as winter is colder and the fact of hotter summers simply results in more precipitation, and therefore more snowfall at high latitudes. It seems instead that the fiercer summers in this hemisphere ensure that ice melts annually, instead of growing into glaciers. Glacial conditions are triggered instead when summer falls at aphelion. These cooler summers may limit the amount of ice that melts at high latitudes, facilitating the growth of ice sheets. Positive feedback then ensures they grow. This corrective to Croll's by then discredited theory was offered in the 1920s, when the Serbian astronomer Milutin Milankovitch proposed similar ideas, in cooperation with meteorologist and the grandfather of plate tectonic theory, Alfred Wegener, and his son-in-law Wladimir Köppen. Counter-intuitively, it seems summer is the key season in the inception of glacial intervals. Milankovitch's ideas fell out of favour in the mid-century as well, to be revived in the late 1960s, underscoring the extent to which new scientific orthodoxies are often built on a foundation of discredited ideas.[18]

'Ice ages' are now understood as falling into two categories, on two different time scales. Long, cool intervals of Earth history, on a scale of tens to hundreds of millions of years, when glaciers wax and wane, are distinguished from shorter periods within these longer intervals, of tens of thousands of years, when glaciers are at their maximal extent. For the sake of convenience, the former are here termed icehouses or glacial super-cycles. The latter are termed glaciations, glacial maxima, or ice ages, and are separated from each other within the icehouses by warmer interglacials. Croll's theories explain the latter more completely than the long interval icehouses. The existence of these super-cycles further explains the deficiencies in his synthesis. Naturally, the bulk of his data related to the most recent glaciation, that of the terminal Pleistocene, which ended about 11 500 years ago, when, indeed, the Northern Hemisphere was icier than the Southern. The existence of cyclic icehouses was barely hinted at. However, one of the groundbreaking implications of Croll's work was the inference, drawn above in the letter from Darwin of November 1868, that the 'N and S glacial deposits' must have 'belonged to distant epochs' and 'the glacial work in New Zealand [and South America] could not have been simultaneous with our glacial period'.[19] This set the scene for the eventual reinterpretation of landforms at Hallett Cove as relics of a far more ancient glacial period than Ralph Tate's hypothesis of 1877 allowed.

The icehouses appear to be concentrated in at least four different intervals, in the Neoproterozoic (between about 850–635 million years ago); the Ordovician–Silurian transition (from 460–430 million years ago); the Permo-Carboniferous (roughly 350–250 million years ago); and the Quaternary (the last 2.6 million years) (see Figure 2). Of these, the latter is the best understood. The last two and a half million years have seen over 20 advances and retreats of continental ice sheets. The causes of glacial super-cycles are understood less well than causes of ice advance and retreat within them. They appear to be the culmination of even longer periods of global cooling, for tens of millions of years before the icehouse commenced. Many factors can produce conditions favourable for icehouses and the inception of 'permanent' ice, from the changing lateral position of the continents (on which, in the absence of plate tectonic theory, neither Croll nor Milankovitch was in a position to speculate), to uplift of continental blocks, to fluctuations in atmospheric carbon dioxide (greenhouse and icehouse effects), to the destabilisation of gas hydrates (linked to sea-level fluctuations), to orbital changes.

The presence of large landmasses at high latitudes is a prerequisite for the formation of glaciers, as such large accumulations of ice cannot form over deep ocean. There is evidence from the Neoproterozoic and from the Permo-Carboniferous icehouses of such super-continents near one or other of the poles. Once the ice begins to form, it becomes a self-perpetuating system, as snow and ice reflect sunlight back into space (a phenomenon called albedo) much more effectively than rock and water. As albedo increases, the planet becomes colder, more ice forms, further increasing albedo in a positive feedback loop. For example, during the Permian icehouse most of the land was concentrated in the Southern Hemisphere, around the South Pole in a giant landmass now known as Pangaea, thereby maximising the effects of albedo, whereas during the later Cretaceous greenhouse the continents had dispersed.[20]

Changes in the lateral position of continents affect ocean circulation. For instance, 100 million years after Pangaea drifted apart, the final separation of Antarctica and Australia created a circumpolar ocean – the Southern Ocean – which isolated the polar continent and may have facilitated the growth of the Antarctic ice cap. Uplift affects atmospheric circulation as well as oceans, as with the great elevation of the Himalayas and the Tibetan Plateau over the last 15 million years. An increase of land at higher altitudes also encourages the build-up of ice and snow.

Carbon dioxide is a greenhouse gas. A reduction in atmospheric CO_2 may cause an overall reduction in the temperature of the atmosphere and the oceans. The amount of the gas in the atmosphere at any time is a result of complex interactions between organisms, ocean currents, erosion and vulcanism, and it is unclear whether the temperature reduction or the CO_2 reduction occurs first in an icehouse. The magnitude of the changes over time is also unclear. However, some commentators suggest that carbon dioxide from volcanoes and submarine vents might have caused a sufficient build-up of atmospheric CO_2 during the proposed snowball Earth scenario of the Neoproterozoic that runaway greenhouse conditions triggered a melting of the ice, before the cycle began again, over tens of millions of years.[21] Snowball Earth theorists also suspect that the destabilisation of gas hydrates may act upon super-cycles. Gas hydrate is a crystalline solid consisting of gas molecules, usually methane. Large amounts of methane and CO_2 are locked up in marine gas hydrate reservoirs. When the sea level drops during a glaciation, these may melt. This has a feedback effect and can prolong super-cycles, or alternatively trigger runaway greenhouse conditions.[22] Variations in the planet's orbit trigger changes in the amount and distribution of solar radiation. This both initiates and affects the course of glaciation. But the cycles of orbital variation are too short to explain icehouse–greenhouse super-cycles and are probably more important for controlling local variation within the larger intervals. It is these more localised changes that Milankovitch's and Croll's theories best explain.

During the 1960s and 1970s, much of this new data generated broad support from the scientific community for Milankovitch's ideas. He had proposed that glacier advance and retreat within glacial super-cycles are mainly due to changes in the amount and distribution of solar radiation reaching the planet's surface. These fluctuations affect the development of continental-scale ice sheets, and result from the same three factors earlier isolated by Croll: orbital eccentricity, axial tilt and the precession of the equinoxes. These all have different periodicities of variation, and changes also vary with latitude and seasons, rendering the composite variations in solar radiation very complex indeed. These recurring fluctuations are now termed Milankovitch Cycles. Data from the last 750 000 years have revealed something of the periodicities and the effects of these three factors in the recent geological past.

Cyclical changes in the eccentricity of Earth's orbit (which varies from between one per cent and five per cent over time) result, as Croll suggested, in seasonal contrast between the Northern and Southern Hemispheres. When the orbit is highly elliptical, one hemisphere will have hot summers and cold winters and the other will have warm summers and cool winters. When the orbit is close to circular, the seasonal temperature contrast will be similar across the hemispheres. The positions and timing of aphelion and perihelion also effect fluctuations in the amount of solar radiation that reaches Earth's surface. These orbital changes have a periodicity of roughly 100 000 years.

The tilt of Earth's axis from the ecliptic – the plane of its orbit – varies from 21.6° to 24.5°. It is currently 23.5°. Over the last 750 000 years, the dominant period of variation has been about 41 000 years. Changes in tilt cause significant changes in the seasonal distribution of radiation at high latitudes, and in the length of winter dark at the poles. It has very little effect at lower latitudes, but with changes in solar radiation of up to 15 per cent at the poles, it has a huge effect on the formation of ice sheets. The timing of the equinoxes changes because the axis of Earth's rotation wobbles. This is known as precession, and it is closely related to axial tilt. Aphelion and perihelion change position on the orbit through a cycle of 360°, with a generalised periodicity of 22 000 years. This affects the seasonal balance of radiation. For example, if aphelion falls in July, summer in the Northern Hemisphere and winter in the Southern will be cooler than the corresponding seasons in opposite hemispheres.[23]

As Croll indicated to Darwin in 1868, the 'grand secret' of 'the' glacial period is the tendency 'to one effect viz the production of snow on the one hemisphere and the melting of it on the other'. This overstates the case somewhat, but still he had a point as far as shorter glacial cycles are concerned. He was unaware of the evidence in Australia for 'arid glaciation', which was not articulated until nearly a century later. It was not so much the 'melting of it', as the absence of precipitation to make snow in south-eastern Australia that produced the familiar semi-arid landforms of the region. The positive feedback effect generated during glacial super-cycles can override the tendency of orbital variation to concentrate glaciers in one hemisphere or the other. Croll cannot be expected to have read these effects in the rocks, as subsequent glacial activity tends to obliterate the signs of previous glaciation, so the cycles of the Quaternary icehouse, and in particular the last few glacial advances, are far more easily recognisable than older super-cycles.

Croll's work was well known in Australian scientific circles by the 1870s. In support of this, in 1879 Gavin Scoular, a wealthy Gawler farmer, an amateur naturalist, member of the Adelaide Philosophical Society and Hallett Cove glaciation naysayer, presented a convoluted account of Croll's work, in which he stated that 'it would be presuming too much ... to enter here in detail upon matters with which a majority of members present are fully acquainted'. Ralph Tate's lecture notes included a fairly detailed Australian summary of some of Croll's ideas.[24] Clearly the acceptance of Croll's theories complicated the interpretation of evidence of Southern Hemisphere glaciation.

Glacial studies in South Australia

Why else was glacial theory important in Australia? Intellectual fashion plays a part. Many late nineteenth-century geologists in Australia had backgrounds in glaciology in Britain or Europe, where the discipline aided in the explanation of changes in past climate and the fossil record. Just as importantly, nineteenth-century glacial studies in Australia demonstrate the intimate relationship between scientific knowledge and subjective experience in the representation of landscapes and landscape processes. For example the Victorian Government Mineralogist during the 1850s, Alfred Selwyn, described experiences in North Wales that enabled him to recognise signs of glaciation in Australia. An obituary for Walter Howchin described the 'glacial hills' which 'produced the mind that later revealed to the world the existence of a glacial age infinitely older than anything previously conceived by man' (the Neoproterozoic icehouse), and Howchin himself described his experience 'of the evidences of extinct glacial action' in the north of England and Scotland. Tate, writing about the geography of South Australia, explained that 'one familiar', as Tate was, 'with the appearance of a glaciated country cannot have failed to recognise a certain resemblance that our hills bear thereto'.[25] European experiences provided the context for the interpretation and imagination of Australian places. So they searched for and imagined signs of geologically recent glaciation in Australia, comparable with the Pleistocene or Great Ice Age of Europe and America, to explain equivalent changes to the landscape and fauna in a different setting. Tate in particular composed a sophisticated synthesis which united extinction, biogeography, human settlement, tectonics and climate change.

Geographic reasons were compelling. A major part of the Australian interior consists of a low peneplain, tectonically relatively quiet with one of the lowest erosion rates known. Weathering profiles dated to the Mesozoic, and correspondingly ancient land surfaces, have survived to the present. The record in the rocks goes way back. In contrast, in Eurasia and the Americas, late-Neogene and Quaternary uplift and glacial erosion produced very different, 'fresher' landscapes. Only in Tasmania and the higher parts of south-eastern Australia are landscapes found which resemble those of western Europe. With its low altitude and low latitude, Australia was not necessarily a fertile garden for the budding glaciologist, but it was a testing ground for uniformitarian principles. So the suggestion of Pleistocene glaciation excited controversy.

The earliest documented identification of glacial landforms in the field was probably by the Reverend W.B. Clarke in the Victorian Alps in 1851 and 1852. He recorded glacial features such as perched blocks – which are erratic boulders embedded in glacial till – during a geological survey commissioned by the Victorian Government. At this time, surveys of the colonies were undertaken essentially with the aim of finding gold. Government geologists frequently expressed their distaste at being employed as mere 'Government Prospectors' and most official observers recorded features of scientific as well as economic interest. From the 1850s to the 1890s, a number of authors proposed that the broad valleys and plains, the lakes and the distinctively rounded hills of southern Australia resulted from the action of glaciers. This was easily refuted and other causal mechanisms were invoked to explain the landforms, such as marine inundations, subsidence, sub-aerial erosion and alluvial activity. As geographer G.S. Griffith noted to the Royal Society of Victoria in 1884, 'we ought not to wonder at rock striae being scarce, but rather we might feel surprised that any should have been preserved ... the smooth-swelling rock surface which tells of massive ice moving slowly across the country and planing down all prominences into flowing outlines'.[26]

Notwithstanding the interest from distinguished metropolitan naturalists, a colonial bureaucratic preoccupation with the hunt for minerals and an early absence of government-sponsored geology probably retarded the systematic search for evidence of glaciation in Australia until the late 1860s. In 1867, in response to an article by Father J.E. Tenison Woods, the New Zealand geologist Julius Haast attempted to draw the attention of Australian geologists to the study of the physical and surface geology of the Australian Alps. Haast wanted to connect evidence of the 'glacier epoch' in Australia with that already established in New Zealand, hopefully thereby to unravel 'some of the causes by which such a remarkable extension ... has been effected'.[27]

Woods' paper, presented to the Royal Society of Victoria, spent some effort refuting the contentious possibility of Australian glaciation. He declared that rather than cold, 'on the contrary, we find evidence of extreme heat, or at least a heat almost tropical in South Australia', and a corresponding 'sub-tropical' fossil fauna. Woods deplored the 'fashion' which posited the 'universality' of a period of cold from which 'even Australia is supposed not to have been exempted'. Haast's gentlemanly opposition to Woods' insistence on the absence of evidence for glaciation in the Southern Hemisphere may have been influenced as much by

Woods' determinedly anti-Darwinian stance as by Haast's own considerable experience of New Zealand's glaciated landscapes.[28] In turn, Haast's mid-nineteenth-century studies in New Zealand disturbed Woods. They compellingly suggested a much greater former extent for New Zealand glaciers during the great Ice Age. Despite his slight reservations, Woods decided that these conditions could only have resulted from exceptional circumstances which did not apply either to Australia or to the rest of the Southern Hemisphere. 'In fact', he wrote, 'the continent is now passing through a colder period than any of which we can find evidence in its previous geological history.' He rejected Joseph Hooker's assessment of 'our alpine flora' as stemming from this 'glacial period of all the world' and lamented that the prevalence of ice 'forms a prominent feature in the graceful theory of Dr Darwin and the speculations of his numerous supporters'.[29]

Woods was avowedly anti-transformist. In 1876, he wrote that he 'had never had cause to entertain any doubt that we are surrounded by species clearly defined in nature, perfect in their organisation'. He concluded that 'Australian geology ... can urge nothing in favour' of evolutionary theory, despite, during two decades of researches in Australian tertiary geology, having 'carefully sought for any reasonable evidence in favour of evolution, or clues to its mode of operation'.[30]

His biographer Ann Player described Woods as 'leaning toward a catastrophist rather than a uniformitarian world view'.[31] However, this is at odds with his rejection of a whole-Earth glacial epoch, as enthusiastically endorsed by the catastrophist Agassiz, in favour of a more Lyellian model. In any case, his statements on the origins of caves have a decidedly gradualist bent: 'Never for ages past enlivened by the busy hum of life', caves 'have within themselves a wondrous record' of the planet's changes. The cavern which 'by the small droppings of water' has 'created itself into a palace', has 'stood silent witness to the earth's history'. While geologists 'have been accused of requiring too much time for the operation of the mutations they have helped to disclose', we need only look upon 'this architecture', this 'glorious tracery of nature' and remember 'that it has been formed atom by atom, line by line'. Considering 'how long it must have taken a mere drop of water to take down from above the marvellous columns which adorn this palace of stone' we must ask whether 'years, even counted by hundreds' could 'cover the period it includes'.[32] Woods corresponded with and greatly admired Lyell, who had commended him on the 1862 *Geological Observations in South Australia*. As a Catholic priest, Woods' position on evolution and human origins, rather than theory of change in the geological record, led him to reject evidence that Darwin had earlier sanctioned, such as the former extent of glaciers into southern latitudes.

In 1867 he wrote: 'If we had no period of extreme cold in the Southern Hemisphere, then the arguments or the theories which account for the glacial epoch, on the hypothesis of changes which affected the whole earth, must be abandoned.' This is not inconsistent with a gradualist approach to geological change, although James Croll's gradualistic models *invoked* 'changes which affect the whole earth' to reject simultaneous glaciation of both hemispheres.[33] Despite Woods' implicit criticism of 'Darwin's hypothesis', less than a year earlier Agassiz had crowed that evidence from the Amazon basin in *favour* of universal glaciation marked 'the end of the Darwin theory'. It could hardly be reasoned that the *refutation* of a 'glacial period of all the world' would be considered a setback by Agassiz' intellectual rival. Croll had provided Darwin with a plausible solution. He allowed for the widespread contemporaneous glaciation of Eurasia and North America, but denied the possibility of a simultaneous period of cold in the Southern Hemisphere. His models also demanded cyclical – i.e. multiple – glaciation. But perhaps in 1867 Woods was not aware of the Scottish geologist's conjectures.

Much of Woods' writing on past climate change seems to consist of *ad hoc* hypotheses. But arguments constructed from eclectic faraway sources tend to be opportunistic and it is demanding a lot to expect consistency in a controversy that spanned more than 20 years. The problems of geographical distribution were as entangled and complex in the mid-nineteenth century as they are at the beginning of the twenty-first. The lack of any unifying doctrine encouraged both this broad-minded opportunism and equal intolerance among naturalists in the selection and rejection of evidence that contributed to such adhockery, and certainly reflects Woods' approach to natural history.

Commonly, evidence for glaciation on mainland Australia was rejected or questioned by naturalists on the grounds that it was too localised to be reliable, and was anyway contradicted by the floral and faunal evidence. Woods based much of his analysis on faunal investigation, recent and fossil, disdaining geological observations which appeared to contradict it: 'One or two instances of grooves or striations are recorded, but standing alone in so vast a territory the ice origin is very doubtful. On the whole, the evidence afforded by the animal remains is decidedly in favour of a warmer climate.' Furthermore 'thousands of fossils and shells' passed through his hands convincing him that 'our continent and seas have passed through a subtropical climate'. The botanical record verified this. The evidence of glacial erratics was equivocal as the transported stones are by definition not *in situ*. They could be attributed to causes other than ice, or if ice-derived, to the action of stray icebergs. So-called 'glacial' striations in bedrock were attributed to tectonic movement and doubters similarly could label 'glacial tills' as 'fault breccia'. Agents other than glaciers which polish rocks include wallabies, tectonic movement, water, stone and 'boys with leather seats to their breeks'.[34]

Curiously, in his *Geological Observations* of 1862, Woods did appeal to glacial mechanisms to explain landforms in the Mount Lofty Ranges: 'Indeed it seemed to me that there were very distinct marks of snow, and the action of glaciers. This would declare the range to have been once of extraordinary elevation, probably the axis of some former continent.' This analysis is consistent with modern plate tectonic syntheses which identify the ranges as distant remnants of the giant trans-Gondwanan Delamerian Mountains of Panel 2. By 1867 he rejected these mechanisms comprehensively, and as late as 1883 wrote that 'There is no satisfactory evidence of any former participation in the great ice age by the Continent of Australia.' Somewhat opportunistically, given Woods' change of heart, Ralph Tate in his seminal 1879 address on Hallett Cove glaciation cited the priest's earlier work and that of his brother, J.D. Woods, in support of past glacier action in the Torrens Gorge.[35]

The first reference to past glaciation in South Australia is probably Alfred Selwyn's description of a striated pavement in the Inman Valley of the southern Mount Lofty Ranges.[36] Lacking one of their own, the South Australian Government borrowed the Victorian Government Mineralogist in 1859, for a whirlwind survey of the colony. During this study, Selwyn discovered a smooth rock pavement with fresh-seeming parallel striations, now known, imaginatively enough, as Glacier Rock or Selwyn's Rock. These grooves he attributed to the action of an ancient glacier, on the basis of observations made years earlier 'in the mountain valleys of North Wales'. This was the first documented evidence for glaciation in South Australia and appeared as an inconspicuous paragraph on page four of his report.[37]

Despite the discovery and his interpretation of it, less than a decade later Selwyn, noting that 'The general absence of mountain lakes of any kind is a remarkable feature in the physical geography of the east Australian chain', asked whether 'this fact may not be in some degree connected with *the concurrent absence of evidence, whether geological or palaeontologic, of a glacial period in Australia*.'[38] Pencilled on the facing page, in government field geologist

Norman Taylor's notebook of 1873–75 which is interleaved with the State Library of Victoria's copy of Selwyn's book is the note 'there is slight evidence'. Taylor then listed

> One siliceous sandstone pebble from the Upper Palaeoz–c conglom. of the Mia Mia (Spring Plains) exhibits traces of groovings. – See also the last page of Daintree's Report on the Geology of the District of Ballan 1866. – Grooved pebbles on the Lerderderg River. – The pebbles of Silurian sandstone in the Lep: [Lepidodendron?] Palaeoz–c beds containing Graptolites.[39]

Although Taylor made no mention of the Inman Valley evidence, Ralph Tate later cited Selwyn's discovery as the earliest evidence for glaciation in South Australia.[40]

As with Woods' about-face, it is interesting to speculate on Selwyn's reason for abandoning a glacial hypothesis. Why did Selwyn ignore evidence he had announced and published just nine years earlier? Perhaps in 1866 he wished to dissociate himself from some of the more dubious claims for recent glaciation such as those Woods rejected in his article of 1867. James Croll's theories were by this time two years in print. Selwyn may have subscribed to Croll's concept of secular summers and winters, which proscribed synchronous Northern and Southern Hemisphere glacial epochs. Or perhaps he thought the evidence was too slight to draw any conclusions.

Hallett Cove and glacial studies

Hallett Cove came to public scientific notice when Ralph Tate announced he had found evidence of 'Late Tertiary' glaciation not long after he took up the post as the inaugural Elder Chair of Natural Sciences at the University of Adelaide. In 1875 or at some time soon after, he travelled 20 kilometres from Adelaide to the cove to gather seashells and was struck by the fresh appearing 'ice-worn' rock faces and boulders on the beach (see Figure 13 below). In May 1877 Tate announced his discovery at one of a series of public lectures on 'The ancient physical geography and geology of South Australia'.[41] His presidential address to the Adelaide Philosophical Society in 1879, soon to become the Royal Society of South Australia, catapulted the issue of recent glaciation of Australia into new prominence. Incontrovertible evidence for Pleistocene glaciation, contemporaneous with the 'Diprotodon beds' of Queensland and South Australia, provided the necessary mechanism for the climate change Tate needed to explain the disappearance of the giant marsupials.

While the apparent freshness of the glacial features led him to date them to the Pliocene, (which we would now term Pleistocene), their low latitude (35° south) and low altitude (sea level) were problematic. He hypothesised that elevation of the region by up to 10 000 feet during the Pleistocene and greater contemporary humidity would be sufficient for the accumulation of snow and ice into glaciers. This is consistent with his transformist views. That Tate had no mechanism for such uplift is largely immaterial, as his major detractors had no mechanisms for the subsidence and marine inundation that they otherwise invoked to explain the features.[42]

Probably the closest Julian Edmund Tenison Woods ever got to the Hallett Cove landforms was in February 1878, when Tate exhibited specimens and photographs at a meeting of the Adelaide Philosophical Society, of which Woods was a member. The first recorded full day public excursion to the site, with the newly formed Field Naturalists' Group of South Australia, was on New Year's Day 1884, when Woods was in Malaysia. On this occasion, nearly 100 ladies and gentlemen 'proceeded by three coaches to Hallett Cove for a whole day excursion' where 'a lively discussion took place', with those 'well-versed in recognising the toolmarks' made by

Figure 13: 13a shows Tate's Rock, Hallett Cove Conservation Park. Rock platforms at Hallett Cove with scratches and grooves identified as the marks of ancient glaciers by Ralph Tate in 1877 (13b shows a close-up view). *(Photograph K. Douglas)*

glaciers 'acquiescing with the opinion' of Professor Tate (Figures 12 and 13). Woods returned from Malaysia but in failing health. He was gradually crippled by paralysis and died in 1889.[43]

Field experience inspired both concurrence and dissent from the eminent attendees. Gavin Scoular was the first to publish his dissent. In 1879 Scoular had published a paper, extensively quoting Croll, which suggested that glaciation could not have occurred in both hemispheres simultaneously, and attributed a marine origin to South Australia's 'drift'. He enlisted Lyell's earlier ice-rafting theory to explain the erratics of Hallett Cove and other parts of the Fleurieu Peninsula. For icebergs to reach the coast of South Australia in a 'secular summer', he required a displacement of the 'frost line' by 8–10° latitude.[44]

Scoular dealt with the Hallett Cove evidence more specifically in a paper in 1885, following several field trips. Unlike Tate, he could not rationalise the low latitude and low altitude with an extensively glaciated landscape in the geologically recent past by appealing to uplift of the land by thousands of feet. The direction of the striations, if glacial in origin, puzzled him and he considered that bedrock weathering, aeolian activity and the erosive effects of water were more plausible agents in such a mature landscape. He also concluded that far from the uplift then subsidence proposed by Tate, the landscapes of South Australia showed signs of subsidence and marine inundation followed by elevation to their present height. Tate responded courteously to Scoular in the same issue of the *Transactions,* submitting that the absence of marine fossils in the South Australian 'drift' argued against its marine origin, that other signs of glacier action were indisputable, and these could not be attributed to icebergs.[45]

Australian drift (called drift in its general nineteenth-century sense of a superficial deposit caused by wind or water) is now known as 'parna'. It is a sheet of aeolian sand and aggregates of clay particles that covers much of arid and semi-arid Australia, in the form of plains or dunefields; it is neither the result of marine inundation nor a remnant of the passage of glaciers, although it may have formed contemporaneously with the terminal Pleistocene glaciers of Europe. Australian geomorphologists consider that parna is the result of extreme aridity during Pleistocene Australia's most recent 'dirt ages', when the dunefields of the arid and semi-arid zones expanded and strong south-westerly winds swept clay and dust across the continent's inland.[46]

The same year – 1885 – in a highly critical synthesis of 'the literature regarding ancient glaciation of Australia', Robert Lendlmayer von Lendenfeld claimed priority for his own discoveries of glacial landforms in the Australian Alps. A physicist and geologist, Lendenfeld attributed these features to a small Southern Hemisphere ice age around two to three thousand years ago. The 27-year-old German mountaineer was clearly not governed by the same laws of civility as Scoular, Tate and Woods. He stated confidently and disdainfully that 'To look for glaciers one must go up to the mountains not down to the sea'. Furthermore, the 'evidence collected by Professor Tate proves by no means that any glaciers had ever existed in Australia'. The Hallett Cove erratics 'were deposited on the beach by ice bergs stranded there'. Finally, he imputed laziness and a lack of rudimentary glaciological sense to his esteemed colleagues: 'Every child in the European Alps knows that glaciers are formed on mountains and nowhere else.' That the 'gentlemen mentioned above never took the trouble to look' for signs of glaciers in alpine valleys indicated 'that of course no reliance whatever can be placed on their statements'.[47]

His confidence paid off and he found, as anticipated, 'most beautiful and indubitable traces of glacial action' in the valleys of the Kosciuszko Plateau. However, his experience of glaciated alpine landscapes did not prevent him from confusing Mount Townsend and Mount Kosciuszko. And notwithstanding his dismissive treatment of Tate's evidence for glaciation at Hallett Cove (which the German zoologist never visited), Lendenfeld assumed, in keeping with Tate's extinction hypothesis, that the Australian glacial period was pluviatile, 'with a dense vegetation in many parts of the country which are now barren, and which was sufficient to feed the gigantic Diprotodon and other fossil marsupials'.[48] Tate does not appear to have responded specifically to Lendenfeld's paper, but Captain F.W. Hutton of the New Zealand Geological Survey dismissed the German's (and also Gavin Scoular's) claim of an iceberg rather than glacial origin for the Hallett Cove erratics. He also remarked pointedly that, without having visited the locality, he could not himself judge of the evidence, 'but the iceberg theory is such a very improbable explanation of the occurrence of erratics in the latitude of Jervis Bay ... that we must hesitate before accepting it as true'. He thought it more probable

> that Mount Kosciusco once stood some three thousand feet higher than at present, when Tasmania was joined to Australia, and Central Australia was, perhaps, a vast lake; than that the temperature of the surrounding ocean should have been reduced ten degrees without any apparent cause, which is the only alternative.[49]

However, by 1887, even Tate considered the possibility – albeit dismissively – that his Hallett Cove glaciation predated the Pleistocene epoch. Delivering the inaugural address ('A century of geological progress in Australia') to the fifth meeting of the Australasian Association for the Advancement of Science (the AAAS) on 26 September 1893, Tate claimed triumphantly – and precipitately – that although

> there have been [geologists] who have opposed and even ridiculed the notion of glaciation in such low latitudes and at such inconsiderable elevations ... today we may congratulate ourselves that a Post-Miocene glacial period occupies an unassailable place in the geological history of Australia. Mr Jack [of the Queensland Geological Survey] has lately added his testimony, as the result of personal inspection, that 'Prof. Tate's observations are correct in every particular' ... This expression of opinion by a master in the art of interpreting glacial signs will, I am sure, carry conviction to the minds of those who till now have been sceptical.

The following Saturday, Tate led 'the largest field trip ever held in the Southern Hemisphere' to the cove, when on the morning of 30 September, '150 ladies and gentlemen took their places on six drays and a wagonette for the contemplated outing' at ten o'clock. A detailed, well-informed, celebratory, report of the excursion was published in the *Observer* of 7 October. Among the learned party was Sir James Hector, Director General of the Geological Survey of New Zealand, who was keenly interested in glacial studies. On this occasion, Walter Howchin first encountered the features.

'Relics of a far more ancient glaciation'

As Tate had predicted, questions as to the glacial origin of the Hallett Cove features were here put to rest.[50] As reported in the *Observer*, there was 'present before them the path of an extinct glacier, on the course of which they were standing, and which could be traced in an almost unbroken line for a distance of two miles along the cliffs'. To the south and west of the party lay the ocean, to the east, stretching northward lay the Mount Lofty Ranges. Hence the 'physical features of the country were such as to suggest that the ice was most likely to travel from the north in a southerly direction'. But the ground was examined and Tate explained his conclusions to the contrary: 'a careful examination of the way in which the ice had planed the rocks over which it moved' showed that 'its passage must have been from south to north', so 'we must assume that in the time of glaciation the high land was to the southward'. Indeed, in the absence of a model such as plate tectonic theory to suggest the lateral movement of continents, these or similar conclusions could hardly be avoided. The putative 'high land' also accounted for the presence of permanent snow at such low altitude and latitude and conveniently matched Tate's hypothesis of post-glacial loss of climatic equability due to the depression of a peneplain as he explained it in 1879.

But it was also at this singular event that questions were raised about the Pleistocene age of the sediments, described by Howchin in 1895. Doubt stemmed from their stratigraphic position in relation to a thin band of fossiliferous sandstone described in Panel 6 (see Figure 7), identified by Tate as 'Miocene', and referred in conventional modern geological histories to an early Pliocene sea. Among others, Sir James Hector expressed a 'strong conviction' that the glacial deposits were older than the sandstone, 'and was probably cretaceous, which would make it synchronous with similar glacial evidences in Australia, and with similar deposits in New Zealand', as reported in the *Observer* of 7 October.

Examination in the field that day proved unhelpful, and the council of the AAAS determined to set aside £20 for the purpose of determining the age of the Hallett Cove deposits. Tate, Howchin and T.W. Edgeworth David were duly appointed. They dug a number of trenches and eventually established that the glacial sediments were overlain by the 'Miocene' marl, proving the greater age of the former. They published their findings in 1895. David and Howchin hypothesised a link with glacigene features from Bacchus Marsh dated to the Permo-Carboniferous. Tate, perhaps protective of his intellectual legacy, seemed less willing to assign the older date: 'The fact that the glacier-path at Hallett's Cove is pre-Miocene should not force us to accept contemporaneity with the glacier-phenomena at Bacchus Marsh, admittedly of Permo-Carboniferous age, which there seems a too-hasty disposition to concede.' In support of an alternative, 'the stratified rocks of New South Wales and New Zealand' indicated other, later glacial periods in the stratigraphic record.[51]

As a result of the interest in the Hallett Cove sediments, the AAAS formed the Australasian Glacial Evidence Committee (1888–1946) in Section C (Geology and Mineralogy), with Tate as foundation secretary. He relinquished the position to David five years later, at the Adelaide meeting of the AAAS in 1893, when Tate was elected President of the association. The committee had a broad canvas in the several decades of its existence, investigating Precambrian to Pleistocene evidence. A New Zealand branch, of which Hutton was a prominent member, sprang up in 1891 to investigate extant glaciers.[52]

At the Brisbane meeting of the AAAS in 1895, David in his presidential address to Section C stated that some of the glacial features of Hallett Cove were probably the work of Antarctic icebergs, and that the glacial beds were all of marine origin. Howchin was quick to discount this in a paper delivered to the Royal Society of South Australia the same year. Though courteously noting that he was 'ignorant of the evidences' David had at his command, Howchin questioned

> how berg ice could polish and grove [sic] hard rocks over extensive areas, whilst maintaining definite lines of erosion and striae, such as we have at Hallett's Cove. I am led to think there is nothing but terrestrial ice, of great thickness, and operating through a long period, that could produce effects such as we have referred to.[53]

The account above of Howchin's paper as delivered to the Royal Society on 3 April and published in the *Register* three days later differs from the version published in the Society's *Transactions*, in that the latter forbore to mention David by name (perhaps as a courtesy to the more senior scientist) and also stated the case against icebergs more strongly:

> In view of the extremely low latitude of Hallett's Cove ... we might have fallen back on the iceberg theory if the features had been reconcilable with such an hypothesis. This, however, is inadmissible.[54]

Howchin went on to correlate the features with glacigene sediments from Bacchus Marsh, dated to the Permo-Carboniferous. They are generally now accepted as being of early Permian or late Carboniferous age, and predominantly of terrestrial, rather than marine, origin.[55]

Far from disappointing, the confirmation of a pre-'Miocene' date for the Hallett Cove glacial features and its correlation with the probable Permo-Carboniferous features at Bacchus Marsh, in southern India, and southern Africa excited most commentators. David wrote in 1895 that 'important as have been the evidences discovered by those who have studied the marks of Pliocene or Post-Pliocene glaciation in Australia and Tasmania, they are dwarfed when placed beside the gigantic relics of a far more ancient glaciation, the true nature of which has been fully recognised only within [a] very recent date'.[56]

Ralph Tate published no more on Hallett Cove after 1895, though he continued to read widely and lecture on glaciation until his death six years later, and by the late 1890s he seems to have accepted the Permo-Carboniferous date, as revisions to his lecture notes suggest. His colleague and former student, Herbert Basedow, took up the cause after Tate's death and continued to plague Howchin for some years, with Howchin commenting at the 1907 Adelaide meeting of the AAAS that 'Mr Basedow practically stands alone in the way in which he has read the rocks.'[57]

It seems Hallett Cove was the making of Walter Howchin. A Primitive Methodist minister, he came to South Australia from England in 1881 aged 36, for the sake of his declining health. He remained a devout Methodist and combined his itinerant preaching with opportunistic field work and teaching about evolution. Eulogised as 'a persuasive expositor of science' rather than a 'bitter partisan', he actively sought to reconcile science and scripture and 'imparted in many hearts the love of God revealed in Nature'. He lectured in Geology and Palaeontology at the University of Adelaide from 1904, as Tate's replacement, until 1920. He continued to publish widely on both glaciation and matters palaeontological, practising as a field geologist until his death in November 1937, aged 92, of complications resulting from heat-stroke incurred while working in St Vincent Gulf during a heat-wave.[58]

Howchin's pioneering work at Sturt Gorge on an earlier Precambrian glaciation is an important part of the Hallett Cove story as well. The Permian glacier that scored and blanketed and froze the Carboniferous palaeo-valley at Hallett Cove carried with it boulders of Sturtian tillite ripped from the walls of Sturt Gorge, formed 400 million years earlier, during the Neoproterozoic, lending its name to what has been described as the most severe ice age in Earth history. Yet another felicitous conjunction of former landscapes is revealed as beneath the crumbling country lie the remnants of another crumbling land.[59]

This leads the Hallett Cove narrative to the mid-twentieth century. Maud McBriar, former lecturer in Geology and patron of the Field Geology Club of South Australia, wrote in a memoir of Geosciences at the University of Adelaide published in 2000 that 'Nothing untoward happened to this quiet spot, then undeveloped, on the rocky coast of Gulf St Vincent' for over half a century. 'And who would ever think of damaging such an esteemed geological site?'

Chapter 3

THE BATTLE FOR HALLETT COVE

In the late 1950s, the location of the cove, 30 kilometres south of the centre of an expanding metropolis, led to the threat of subdivision and the destruction or burial of the historically interesting glacial pavement and other features considered important in the story of the assembling of South Australia. On the 1962 State Development Plan for the metropolitan area of Adelaide the cove was rezoned from rural to residential, which made the issue of its protection ever more pressing. Nobody initially raised any formal or systematic objection to the rezoning, because the report was the first of its kind in the state and most South Australians had little appreciation of its implications. As a consequence, when subdivision plans became public, landholders planning to develop and groups interested in preserving the natural environment and amenity of Hallett Cove began a series of campaigns that eventually spanned 20 years. These disputes loosely divide into three phases, the first from 1957–1965, the second from 1965–1969 and the third from 1971–1976.

In 1957 the National Trust of South Australia began negotiations to preserve a small 'representative' strip of coast including Tate's striated pavement (Figure 13), which culminated in the dedication of the three-hectare Sandison Reserve in 1965. From 1965, Maud McBriar and other staff at the University of Adelaide, particularly from the Geology, Botany and Zoology departments, members of the Geological Society and the Nature Preservation Committee of the National Trust of South Australia, mobilised an eclectic and formidable array of Australian and international scholars, students, heritage bodies, politicians, school teachers, naturalists and local communities to oppose successive attempts to subdivide and develop the cove.

In 1969 they succeeded in gaining limited protection for 20.6 hectares (51 acres) of cove land. This area was reserved as a site of scientific interest (or SSI) under new South Australian Government legislation (Sections 36 and 38 of the *South Australian Planning and Development Act, 1966–67*). The boundary, selected by the Director of Mines, failed to encompass many features considered to be of scientific importance. Furthermore, under the Act the title remained the property of developers who had purchased the land early in 1969, after reservation had been mooted in 1967.[1]

In 1971, the campaigners demanded major open space legislation recognising Hallett Cove as an outdoor laboratory and museum. This depended upon state government purchase and expansion of the site of scientific interest to include a buffer zone. In 1972 the state government, 'prompted by overwhelming public concern' and acting 'to what is believed to be the limits of its financial resources', acquired 47.75 hectares (118 acres) of land adjacent to the Sandison Reserve.[2]

Further campaigning between 1972 and 1975 eventually secured Federal government intervention. Under the recently established National Estate Program, the Federal Minister for Urban and Regional Development, Tom Uren, commissioned an investigation of the 'scientific, environmental and social issues relating to urban development' at Hallett Cove, known as the Rudman Report (P. Rudman, P.W. Cullen, K. Enright, C.D. Ollier and R.W. Robertson, 1975, *Hallett Cove Study Report: An investigation of the scientific, environmental and social issues relating to urban development in the Hallett Cove area*. Australian Government, Canberra). With subdivisions pressing upon its hems, a conservation park of approximately 50 hectares (some 123 acres) was gazetted on 1 July 1976, to the qualified satisfaction of the groups and individuals concerned with the preservation of the natural environment at Hallett Cove.[3]

Changing priorities in natural heritage management

The terms of the debate shifted as the voices of conservation and preservation gained cohesion and confidence throughout the 1960s and 1970s. Among other factors, the campaigners' escalating demands for land to be reserved reflected a new articulation of notions of natural heritage and national estate, growing public interest in environmental protection, a changing demographic and consequently shifting priorities within the heritage movement nationally and an increased demand for public land, public consultation and bureaucratic accountability. The campaigns of the 1950s established the right of scientists to 'speak on behalf of nature'. As Libby Robin has shown, during the 1960s, radical ecology widened the sphere of authority with demands for 'cultural and aesthetic arguments, as well as democratic participation'.[4]

In 1957, the National Trust sought to purchase only a small section of land – Tate's Rock – the importance of which lay in its perceived embodiment of a moment in the intellectual history of South Australia, when Ralph Tate discovered signs of its past glaciation. Between 1965 and 1969, the Trust, the Geological Society of Australia and other interested organisations campaigned for the preservation of another relatively small piece of land adjacent to the Sandison Reserve, 20.7 hectares (51.25 acres), barely half the recommended area. In 1975, Warren Bonython, G.W. Toogood, Maud McBriar and Ifor Thomas in their capacities, respectively, as President, Director, Chairman of the Nature Preservation Committee and Member and past Chairman of Nature Preservation Committee of the National Trust of South Australia retrospectively defended and explained the Trust's policy in a submission to the Rudman Committee of Enquiry. They pointed out that the Trust believed that 'any land which could be added to the Sandison Reserve would have to be acquired by purchase, eventually'. So, given its limited financial resources, the organisation supported 'schemes which proposed minimal extension of the boundaries to conserve the most important of the scientific features'. Accordingly, decisions on boundaries 'were frequently determined by economics and by the tactics of settling questions with one land owner at a time' rather than seeking to reserve the entire area of scientific and historical significance. In defence of the 'pragmatic approach', they wrote that 'gains were made'. But in the early 1970s, in conjunction with a number of other organisations and individuals and in the wake of public and media outrage at perceptions of the developers' intransigence, greed and double-dealing, as well as imputations of local government opacity, incompetence, corruption and high-handedness, the National Trust threw such caution to the winds. It demanded the state government intervene to prevent further damage to the site of scientific interest and to important geological, archaeological, botanical and historical features that fell outside the protected area.

Preserving the broad panorama

This shift in policy roughly coincided with practical changes in the way in which Hallett Cove was studied and valued by geologists, associated with the growing acceptance of plate tectonic theory as a way to understand the shape of the world, as McBriar explained in 1975 in a submission to the Rudman Report on behalf of the Geological Society:

> For explanations of geological concepts of an ancient ice-covered land mass with glaciers coming down to the sea, continental drift and plate tectonics, *a broad panorama is desirable.* This will be destroyed if houses are permitted to overshadow the reserve. The shift in emphasis in geology from an earlier interest, primarily in an ancient glacier, to the modern concern with wider aspects has led geologists to realise that more than a few acres of reserve is needed to preserve and to demonstrate the totality of the geology exposed here.

Hampered by lack of funds and unfavourable legislation, the National Trust nonetheless continued to press all levels of government and attempt to engage public interest. Supporters were gratified when the South Australian Government acquired 47.75 hectares (118 acres) of cove land for a reserve in 1972 and 1973, although the boundary fell short of the area the Trust and its allies had campaigned to protect.

The Trust's change of policy also reflected the need for public space outside the reserve, in the form of a buffer zone, to protect the features as much as possible from idle curiosity, recreation, 'rockhounds, gardeners, and boat-owners' and excessive 'trampling'.[5] Dr Ron Caldicott, the Director of Research at the Town and Country Planning Association of South Australia, was also an Executive Member of the Australian Conservation Foundation. In a number of articles written and statements made between 1971 and 1975, he harnessed awareness of this need for Major Open Space legislation, both for improved quality of urban life and to protect the conservation park, to a rhetoric of loss and a social equity agenda:

> a major district open space should be declared ... bounded by the railway on the east, the Hallett Cove road on the south, and the existing development at the northern edge ... The buffer zone created by ... this open space would have taken the impact of people off the beach and thus the scientific areas would be protected. Undoubtedly, the recreational effects of a community of thousands of people would inevitably destroy an irreplaceable asset ... the impact ... on such an area of real scientific and historical significance will be disastrous ... regardless of what efforts can be made to protect the reserve by the National Trust or any other body. ... attention should have been given to the needs of society as a whole as well as those of the subdividers. The atmosphere of Hallett Cove has been lost forever. We will have suburbia, with minimum size allotments near Hallett Cove which will never be the same.

By 1975 the Trust's earlier more conservative policy was seen as naive. In support of 'many conservationists and other bodies in this State', it demanded that the boundaries of the Hallett Cove reserve be extended to the railway line to the east and Grand Central Avenue to the south. In its role championing 'causes such as this which will assist in the preservation of the national heritage', it understood the action at Hallett Cove to be 'the last opportunity to act before the land is settled with houses and beyond redemption':

> This is an area which is priceless in terms of its geological contents and, if destroyed or further seriously mutilated, would spell the end to the effective presentation of processes that have taken more than 600 million years to be created. The Trust strongly advocates protection of the total area described in this submission.[6]

The campaign to save or develop Hallett Cove between 1960 and 1976 involved three layers of government, real estate developers, academics, the mainstream Adelaide media, teachers, heritage practitioners and locals and a changing understanding of places worth saving. It divides into three significant stages, discussed below.

1957–1965: Professor Alderman writes to the National Trust, which spends eight years negotiating with a local council to preserve 'a few scratches'

Arthur Richard Alderman, Chair of Geology and Mineralogy at the University of Adelaide, made initial moves to protect some of the cove in 1957. He approached the City of Marion Council, who advised him to seek out the relevant landowner, George Sandison, and suggested that he 'report to the National Trust that there are many places of geographical interest in Halletts Cove, with the object in view of the National Trust taking over some portion of Halletts Cove property as a Public reserve'.[7]

On 5 April, Alderman wrote to the newly formed Nature Preservation Committee of the National Trust of South Australia. The Secretary, H.G.C. Kempe, and the Chairman, Norman Tindale, inspected the Hallett Cove land on 5 August and 29 August 1957 on behalf of the committee, which then commenced negotiations with George Sandison. Sandison agreed in principle to donate three hectares (7.4 acres) of land to the National Trust, containing Tate's Rock, but Sandison died in 1958 well before negotiations were complete. This slowed proceedings, as the Trust then had to negotiate with Elders Trustee and Executor Company Ltd, acting on behalf of Sandison's 30 beneficiaries.

Initial investigation by Elders Trustee revealed that there was a surveyed but impracticable government road along the whole sea frontage of Sandison's land, and that it was impossible to identify the boundary between the Sandison Estate and government land. In 1959 the Minister of Roads informed the Trust that a survey would be necessary to 'establish if the pavements [were] on the road reserve', and that the land was vested in the Corporation of Marion, which should be able to indicate both the actual location of the pavements in relation to Sandison's land and whether there was to be a road constructed in the area.

On meeting with Nature Preservation Committee members Warren Bonython and Jeffrey Clarke in April, representatives of the Marion Corporation stated that there was no road along the cliff top as indicated by the Minister of Roads, and that the glacial pavements were within the 150 links coastal reserve and therefore in the South Australian Harbours Board property. They informed the Trust that the corporation had no jurisdiction in the matter. However, in June that year, a survey of the area was carried out which indicated that the Minister's statement regarding the existence of a road, under control of the Marion Corporation, was correct.

In November 1959, Elders Trustee informed the National Trust that Sandison's beneficiaries had agreed to donate a strip of land one chain wide alongside the segment of surveyed road containing the glacial pavement, to be called the Sandison Reserve. This was despite strong recommendations by Professors Alderman and Rudd and Dr A.W. Kleeman, all from the University of Adelaide, that a far larger area should be preserved. But Bonython and other members of the Nature Preservation Committee were afraid that if they requested a larger boundary, they might prejudice the Trust's chances of acquiring any of the land.

Negotiations with the City of Marion Council continued over the next five years as the Trust endeavoured to obtain the strip of road abutting the reserve, on the cliff top above the beach and tidal flat. Meanwhile Marion Council attempted to force the Trust to negotiate on the council's behalf with Sandison's beneficiaries, to encourage the latter to donate to the council an equivalent portion of land to that requested by the Trust. For example, Jeffrey Clarke noted that at a meeting on 14 December 1959, the council approved the closure of the coast roadway provided that 'a similar strip of land to the east in lieu thereof is made available for roadmaking purposes'. In further consultation in January 1960, Marion Council announced

Figure 14: 'The area was most uninteresting.' Awareness of the deep past at a place like Hallett Cove, as revealed in its sediments and features like the perched block seen eroding from the cliff in the centre of the image, can transform perceptions of its scenic beauty and worth. *(Photograph K. Douglas)*

that the Trust must approach Sandison's beneficiaries for additional land for a substitute road. This included the unsubstantiated threat that the state government was considering the construction of a coast road 'within the next twelve months' which would cut across the unprotected glacigenic sediments and features and seriously compromise the Sandison Reserve. The Trust responded in January 1960 that it was 'financially and practically impossible' to construct the road along the present survey, which in places overhung sheer cliffs, up to 30 metres above a boulder-strewn beach. Under these circumstances the Trust 'felt that the closing of the road would cause little or no inconvenience to the council and certainly no extra cost'.[8]

Prominent members of the University of Adelaide Geology Department, the Royal Geographical Society, the Geological Society of Australia (SA Division), the Royal Society of South Australia and the National Trust continued to press for the closure of the road, sending deputations to the Minister of Roads and Marion Council. On 10 October 1962 the council advised that approval had been given. Further delay transpired when Ruby Coombs, owner of 'The Leys', which bordered the road, objected to its closure on the grounds that 'she thought the area was most uninteresting'.

The reserve was expanded to 5.5 hectares (13.5 acres) in 1965 when the Lands Titles Office transferred the title from the Marion Council to the National Trust. Its official opening was held on 20 November. Clarke, who was secretary of the Nature Preservation Committee during the greater part of the negotiations, attributed the five-year delay to 'continued procrastination by the Marion Councillors and Council officials', who, like Miss Coombs, may have failed to see the worth of a few scratches on a windswept pavement atop a scrubby cliff (see Figure 14). Meanwhile, the Field Naturalists' Society of South Australia had been lobbying the National Trust since mid-1960 for protection for land bordering the Amphitheatre south of Black Rock, owned by Woodend Park Ltd. Following the successful acquisition of the Sandison Reserve, the Trust's attention turned to this matter.[9]

1965–1969: Maud McBriar challenges the powers of monopoly capitalism at Hallett Cove and strikes a blow for geology

McBriar has said that a major element that emerged from the 'Save Hallett Cove' campaigns was a growing awareness from geologists that they could not afford to be complacent about matters and sites of geological heritage. Without an educated public, the value and interest of many geological sites is far from clear. One woman's historic glacial pavement is another woman's 'most uninteresting' beach frontage. At the same time, the lack of public education about sites of geological importance, as well as lack of clarity in the assertions of value for the site by heritage professionals, Earth scientists, botanists and other campaigners, and the Trust's early cautious approach to reserving land, left local politicians, real estate developers, urban planners and consultants somewhat in the dark about sites worth saving.

In 1965, McBriar was a senior demonstrator in the Geology Department at the University of Adelaide. She taught an extra-mural course to regional planning students at the South Australian Institute of Technology (SAIT). At her suggestion, they began a project on Hallett Cove and came to her with the 'alarming news' that a dense housing subdivision had been drawn up for land bordering the newly dedicated Sandison Reserve. As McBriar has subsequently written, this threatened to destroy features of the coastal hinterland that were 'fundamental' to the geological story of the area. They wrote a report proposing a 40.46 hectare (100 acre) conservation park to preserve areas of historic and scientific interest. The School of Architecture and Building at SAIT granted permission for its widespread circulation. Graham Whitten, Chairman of the South Australian Division of the GSA and Chief Geologist in the South Australian Department of Mines gave his support in a letter to accompany the report. They received responses from the South Australian Premier, Francis Henry Walsh, from Don Dunstan, then Attorney General, and from Sir Thomas Playford, Leader of the Opposition, an indication that the conservation of urban land was an issue capable of rousing the interest of the state's political elite.

McBriar's involvement continued when Ifor Thomas, then acting Chairman of the Nature Preservation Committee of the National Trust and a zoologist at the University of Adelaide, invited her to attend an offical 'viewing' of the terrain 'with the object of delimiting as far as possible what needs to be preserved' on 2 September 1966. Those in attendance included Thomas, McBriar, two other members of the National Trust, including Charles Alexander who was there in his capacity as representative of his firm of surveyors, who had prepared subdivision plans for the potential buyer (real estate developer K.J. Powell), the Director of Planning at the State Planning Office, Stuart Hart, and Professor Martin Glaessner of the University of Adelaide as 'an expert witness on geology'. Neither representatives of Powell's company, which eventually purchased land from Elders Trustee, nor landowner J.S. Sheidow attended, though both had been invited.

Thirty-two hectares (80 acres) of the land fell within Powell's property 'and covered most of the geological features' and Sheidow owned 8 hectares (20 acres) to the south. Continuing its policy during the Sandison negotiations, the Trust again asked for considerably less land than they or the GSA considered necessary to protect or represent the areas of geological, archaeological and historical interest.[10] They justified this pragmatically, on the basis that the land belonging to Powell was threatened more immediately by subdivision and that 20 hectares (50 acres) of this 'was the absolute minimum area for the preservation of the essential geology and of one of the more important of the aboriginal campsite areas'. Negotiations were simpler for involving only one landowner. Economic reasons governed the decision, as the Trust feared that it or a similarly cash-strapped organisation would need to purchase the land to ensure its preservation and in 1967 Powell valued the 10 hectares (24 acres) regarded as having greatest geological significance at $250 000. Adelaide architect Dean Berry, then president of the National Trust, pointed out in a letter to Don Dunstan, who was by then the South Australian Premier, that 'this can only be regarded as a gross over-estimate of its real value either for land development or for its mineral potential'. But the Trust was in the invidious position of trying to put a price on perpetuity in a seller's market.

Although aware that the SSI legislation could impose restrictions on development, Western Australian developers Kadima and Silesia eventually purchased the parcel of Powell's land incorporating the site of scientific interest. In a development application of 1971, the companies flagged an inconsistency that resulted from the Trust's policy of asking for less land than protected the important geological features of the cove. Although cognisant of the scientific arguments for preservation of the area, the companies were 'unable to find a consolidated case as to why the boundaries of the reserve were drawn as they were'. Accordingly they sought 'independent' scientific appraisal from a Perth-based consultant, T.S. Martin and Associates, to satisfy themselves of 'the scientific merits of the gazetted boundaries' and to make a case for 'rational definition' of the SSI based on investigation by 'a responsible scientific organisation'. The consultant's report concluded that the land most suited to high value urban development, although alienated from the companies by the reservation boundaries, did 'not show any features of geological significance'. Moreover, archaeological sites north of Waterfall Creek did 'not warrant ... preservation as Sites of Scientific Interest'. Further apparent inconsistency pointing to 'arbitrary definition' was evident in the treatment of the Amphitheatre:

> If the formation is of such importance why should it be reserved in part only and only on one owners [sic] land? ... Why should one owner of land not be allowed to disturb the amphitheatre when a government department recommends major earth-moving on an adjoining part of the amphitheatre on another developer's property?[11]

The application questioned the value of land which had already suffered 'considerable depredation from humans' and concluded that 'the present definition of Regulations Area boundaries is inconsistent, arbitrary and discriminatory as applied to land held by the Companies and scientific evidence is evinced to prove that boundaries should be redefined'.

Opponents countered that a Western Australian company which specialised in marine science was not suitably qualified to judge the geological and archaeological merit of a canonical South Australian site. Kadima and Silesia were unsuccessful in their application to develop six hectares (15 acres) of land within the perimeter of the SSI. Indeed, the main effect of the report, and ensuing public outrage, was the state government's decision to acquire the land in 1972, and move to protect more of the land outside the SSI. But the consultants' report underscored the confusion engendered by the legislation and by the specification of the boundaries.

However valid the Trust's subsequent rationalisations, its policy mystified supporters and opponents alike, and the SSI regulations provided only limited legislative protection for specific features. But geological heritage and legislative protection for sites of scientific importance were fairly unfamiliar territory in 1965. It was a steep learning curve for both scientists unseasoned in the 'destructive, exploitative', 'rapacious' ways of urban development and for would-be subdividers unaccustomed to the 'protracted manoeuvres' and 'emotional political opportunism' of groups such as the National Trust and the Geological Society.[12]

McBriar helped stir a mixed brew of crusaders – scientific and lay, individuals and organisations, from home and abroad, political and community groups, with broad agendas and eclectic political affiliations. This mix continued into the final stage of the campaign. Evidence for this is available from a number of sources, including letters of support from interstate and international geologists; notes on the Third Annual Gondwanaland Symposium which held a field trip at the cove; a steady stream of letters to the editor in the *Advertiser* from 1966 to 1976; and the eventual response of local, state and federal governments. From 1965, sympathetic coverage sporadically featured in the *Advertiser* and other Adelaide papers and several feature writers, most notably Stuart Cockburn, adopted the cause as their own.

1971–1976: The Hallett Cove Development Company misjudges the tide of public opinion and gets caught in the rip

In 1970 the Hallett Cove Development Company proposed to develop land to the south of the SSI, within the Amphitheatre, as a marina and boat haven. The company planned to acquire material by quarrying stone from the same area. Kadima and Silesia proposed to develop six hectares (15 acres) of the SSI as prime beachfront residence at the same time. A media hubbub ensued when news of the plans were leaked in September 1971.

While heritage bodies, University of Adelaide geologists and the Geological Society were at the vanguard of the earlier campaigns, 1971 saw an unprecedented level of grass-roots support and partisan media participation as the *Advertiser* increased its promotion of the 'Save Hallett Cove' movement. Between 15 and 22 September 1971, at least 20 articles, cartoons, editorials and letters about Hallett Cove appeared in the *Advertiser*, and several more appeared in the *Sunday Telegraph* and the *News*.

Members of the National Trust, Geological Society, Field Naturalists' Club, University of Adelaide, the Town and Country Planning Association and state MPs such as the Member for Kingston, Dr Richie Gun, and Don Hopgood (later State Minister for Environment and Planning) continued to play prominent roles. But the public campaign was spearheaded at a local level by Marion Shire residents like Cumberland Park garage manufacturer R.S. Alcock and Morphettville mother-of-four Jean Minards, respectively the president and the media liaison officer of the Save Hallett Cove Committee. As McBriar recalled in 2000,

> Many organisations supported extended conservation [including] a spontaneously formed local group of citizens constituting the 'Save Hallett Cove Committee', students from Adelaide University and from other tertiary institutions, secondary and primary schools. Support for the campaign was overwhelming and gratifying – it came from all parts of the State and from as far afield as the UK.

On 10 September 1974 Gun, who had received several representations from constituents, invited Moss Cass, the Federal Minister for Environment and Conservation, to visit the area, in company with McBriar as a representative of the newly founded Field Geology Club, and other representatives from SASTA, the Marion Council and the state government. Following this excursion, Gun informed SASTA that Cass had 'undertaken to discuss the matter' with Tom Uren. Uren indicated his interest and support in a letter to Gun of 18 September. Cass announced on ABC radio on 10 September that 'it was the Australian Government's responsibility to preserve areas of significance at Hallett Cove'.

Sadly for supporters of conservation at the cove, their applications for National Estate money for Hallett Cove were not granted, due in part to confusion arising from the National Trust's earlier cautious policy in delineating the SSI boundaries. But in 1974 the Hallett Cove Development Company graded the south side of the amphitheatre. This resulted in Federal Government intervention, a change in legislation to prevent such future action by developers, the commissioning of the Rudman Report and the eventual gazetting of the park's boundaries in 1975, which is not the result the company anticipated when it brought in the bulldozers.

Geological heritage and the outdoor laboratory

Another historical glaciological inquiry provided some of the impetus in the campaign to preserve the Kosciuszko Tops in the 1950s, fought by A.B. Costin and the Alpine Ecology Unit of the CSIRO and by the Academy of Science. As Libby Robin has written, these early sites of scientific controversy are as much a part of Australia's intellectual history as the countryside itself is part of its aesthetic, scientific and economic heritage. The rhetoric which insisted on their importance, and of the area in general as a site of ongoing research into glaciation in Australia, provides a strong resonance with the Hallett Cove story. The smattering of iconic scientific surnames which label the landforms of the area even evokes some of the same personalities, notably David Moraine, after Edgeworth David. Another echo is the campaigners' claim to an invaluable teaching resource, an outdoor classroom or laboratory (although from a botanical as much a geological perspective). So the Kosciuszko Tops debate was a possible precedent for Hallett Cove and many parallels emerge, but it does not seem overtly to have influenced the scientific campaigners. The Hallett Cove campaigners instead connect their genealogy with the efforts of the Nature Conservancy movement in the United Kingdom in the 1930s and 1940s. The last phase of the Hallett Cove campaign bears comparison with the near-contemporary Little Desert campaign in Victoria (1968–1971), the focus of Robin's important 1998 study of the rise of ecological consciousness in Australia, and itself illustrative of a paradigm shift in the awareness of places worth saving.[13]

The modern conservation park is commensurate with such sites of geological and scientific heritage as Lyme Regis in Dorset, the setting for part of Jane Austen's novel *Persuasion*, best known for the World Heritage listed lower Jurassic marine fossils which abound in its cliffs, and Wren's Nest National Nature reserve in Dudley, Worcestershire, which is renowned as both a Silurian fossil site and also as a monument to the Industrial Revolution. The National Trust and the Geological Society of Australia have initiated a 'glacier trail' linking several sites along the Fleurieu Peninsula, such as Rosetta Point, the Inman Valley (site of Selwyn's Rock) and Hallett Cove. It now extends 1200 kilometres from Cape Jervis on the Fleurieu Peninsula to Parachilna in the Flinders Ranges. A similar project in Wisconsin, run by the Ice Age Trail Alliance, manages a 'thousand-mile footpath' entirely within the state, which links sites of a more recent glaciation, that of the terminal Pleistocene. The alliance seeks to tie sites of geological importance to the places of natural beauty where they occur, working with the Wisconsin Department of Natural Resources, National Park Service, local governments, businesses and private landowners, an undertaking which resonates with the Fleurieu Peninsula trail.[14]

Colin Michael Hall wrote in 1992 that the 'creation or continued protection of a national park or wilderness area is not a rational process. It is a political battle.' While not an untouched wilderness in any conventional sense, the rhetoric of the Hallett Cove campaigns is steeped in the same language of struggle as efforts to conserve emblematic wilderness areas like Tasmania's Lake Pedder or Queensland's Fraser Island a decade later. A vocabulary of primitive areas and wilderness was in any case invoked in depictions of the cove. Campaigners described it as primitive, in a natural state; expressed fears that 'the entire landscape will be domesticated and "prettified", and its present character will be erased'; insisted that changes would be 'as disastrous to the wonderful inshore natural formations', and hence to Adelaide's character, as would be the conversion of other iconic South Australian wild places like the Flinders Ranges' Wilpena Pound 'to a new respectable suburb'. Although undoubtedly a cultivated landscape, its rugged, eroding cliffs, badlands features and scrubby vegetation place the cove somewhere between domestication and wildness. Standing in place of a land surface unmodified by human activity, the cove's geology evokes an imagined wilderness of the deep past.[15]

Activists and environmental historians Drew Hutton and Libby Connors have labelled the campaigns to prevent development at Hallett Cove 'the first step towards the invigoration of community environmentalism in South Australia'. Prominent supporters from the Adelaide establishment and scientific organisations included the Royal Society of South Australia, the Australian and New Zealand Association for the Advancement of Science (ANZAAS), the National Trust, the Anthropological Society of South Australia, the Geological Society of Australia, the Australian Academy of Science, the South Australian Science Teachers' Association, and a number of politicians including Premier Don Dunstan. But despite the establishment and scientific credentials of many of those who lobbied to protect Hallett Cove, the campaign has been constructed, both retrospectively and in the press reports of the time, as a crusade against exploitation (as opposed to rational, environmentally sensitive urban planning), as a recognition of the power of the people to combat local government corruption and high-handedness, as a victory for public science education and the heritage movement in South Australia, as much as a vindication of the significance of geological heritage.

Comments by prominent South Australian and Federal politicians in newspapers and press releases emphasise this. Dunstan announced in a press release on 14 October 1971 that he had decided that 'in fairness to the public and developers that now is the time to announce that there are some forms of development this Government will not accept'. Reflecting a shift in focus from the protection of an intellectual and historical resource to averting a social and environmental crisis, he affirmed that development at the cove 'is completely unacceptable from the viewpoint of public amenity'. Reports on the press release continued that Dunstan noted that 'the uttering of words like "development" and "progress" by developers did not and should not automatically confer some kind of blessing upon their plans, however much they may be worth' to the state. As 'a vital property of the people of South Australia', Dunstan promised that his government would protect such places: 'There must come a point at which all Governments in Australia take a stand to protect natural resources, and we are taking a stand at Hallett Cove because the project, though superficially attractive, is socially disastrous.'

Environmental campaigners characterised the developers as Machiavellian, selfish, underhand, 'questionable', 'based on the gullibility of the public' and their actions as 'pillaging', 'exploitative' and 'desecration'. The early protective measures of state government ministers and departments, where considered to fall short, were categorised as 'tokenistic', 'short-sighted', 'lacking vision', and 'minimal'. The literature of conservationists likened processes of urbanisation to the spread of a disease or an alien terror, mirroring the language of Cold War science fiction/horror films, as the 'tentacles of urban sprawl' spread across the state encroaching on the last refuges. Pre-development Hallett Cove was a botanical refuge, 'close to its natural condition', a geological Mecca, an invaluable resource, 'a beauty spot', a picturesque landscape, 'a source of beauty and pleasure', a living science lesson and 'St Vincent Gulf's most beautiful natural cove'.

In contrast, the vocabulary of developers, while acknowledging the 'magnificence of the outlook' and the 'panoramic views' as assets, stressed the 'deficiency' of the beach, the lack of scenic attraction of the 'steep and broken landform', the 'rapid degeneration' of the area which has been 'defaced', 'denuded' and 'degraded', not by subdivision but by agriculture and by 'uncontrolled public access' resulting from the site's isolation, to which sensitive urban planning could provide a corrective. For an opposing argument, of the unsuitability on pragmatic grounds of the site for residential subdivision, the SAIT report prepared by Maud McBriar's students, which precipitated stage two of the conservation battle, pointed out that 'if the subdivided area had particular merit for a residential area, its reservation might be thought more contentious. But on the contrary, it is steep and exposed to the full force of western sun and weather.' Libby Robin has pointed to similar arguments about the lack of viability for intensive settlement of particular areas of north-western Victoria which were made of Sir William McDonald's Little Desert Settlement Scheme during the campaign to save that area in 1968.[16]

The Hallett Cove campaigners' language evokes a prelapsarian wilderness: 'nature in its untamed state' – 'magnificent and unspoilt'. In a number of articles, submissions and letters to editors, supporters were asked to consider 'the whole experience of the wind on the hills, the loneliness of the deserted beach, the long stretches of nature undefiled and timeless' and the 'stark and rugged character of the hills above the cliffs' of the cove which, while 'close to its natural condition', was 'a source of beauty and pleasure'. Subsequent development had 'destroyed the whole character' and left the 'stark, rugged character' blurred and dissipated. Adelaide resident and Professor of Meteorology at Flinders University, Peter Schwerdtfeger, wrote to the *Advertiser* in July 1974, describing the grading of the terrain above the beach as 'an act of total naivety and irresponsibility', deprecating the 'brutally obvious railway embankment' which further debased the 'sordid scene' and lamenting that 'the foreshore of Hallett Cove is gone and SA has allowed yet another unique landmark to be senselessly scraped off the map'.

Much of this rhetoric is patently at odds with a Hallett Cove landscape which had been grazed, mined and farmed for well over 100 years, as pointed out by spokesmen for the development companies who, in a letter of 23 July 1971 to Glen Broomhill, the State Minister for Conservation and the Environment, declared that 'there has been some emotional thinking on the subject' and accused conservationists of opportunism. It was also disingenuously heedless of evidence for pre-European habitation and modification of the landscape in the form of tool factories and fires reported by Matthew Flinders in 1802 as he sailed up Gulf St Vincent. However, any embellishment in the statements of the campaigners is tempered by legitimate grief at the perceived or imminent loss to the South Australian public of 'yet another unique landmark' and anger at the bureaucratic opacity, hypocrisy and lack of consultation that leads to such 'decay'. Writers pointed out that, although cognisant that responsible urban development can play a role in conservation, a nation 'falls into decay when its standards are so lowered that only material benefits are of account'. Following the grading of the Amphitheatre, 'the worst predictions' of conservationists and scientists 'who have fought for years to preserve the area against close development' had been realised. As an editorial in the *Advertiser* claimed in September 1971,

> Once again, a local council, this time the Marion Council, appears to have abandoned the interests of the environment ... Surely Adelaide has enough subdivided, bituminised, despoiled and ruined beaches without adding another? And surely now is the time to arrest the development which is making almost the entire coast ... into one unattractive and undesirable conglomeration? ... the Marion Council's approval of quarrying at the Cove has made one point, already evident at Stirling, very clear. As part of the State's heritage, the best of the environment cannot be left to the hands of local government.

If the general lack of topographic loftiness west of the Great Dividing Range effected a local understanding of places worth saving in South Australia at odds with the monumentalism which sometimes underpins the appreciation of younger landscapes in tectonically active settings, the message was no less one of grandeur. In the 1950s wilderness was a 'hot political issue'. Robin, writing about the Little Desert campaign, stated that the Little Desert National Park was seen as a 'monument to lost wilderness, an element of the past retained for the future ... It was the holistic ideal ... that conservationists sought to defend.' Likewise the campaign literature of Hallett Cove insists on the need to preserve the site, with its 'almost unbelievable glimpses of the past', for 'the future'.[17]

If, as Hall suggested in 1992, South Australia's geographical circumstances required a system 'that could overcome the paucity of snow-clad mountain-top scenery', then the immensity of deep time in the intimate setting of Hallett Cove provided just that. Its features appeal to a dynamic, vernacular Australian landscape aesthetic incorporating a sense of time depth, vastness and the evocation of change through time rather than space. For example, in 1977 the South Australian division of the GSA nominated Hallett Cove to the Register for the National Estate. The nomination document describes the glacial features as 'a world renowned area of geology', the 'best of the (late Palaeozoic) Australian glacial pavements ... amongst the best in the world'. The park boundaries protect a 'world renowned area of geology' preserving the glacial pavements 'in relation to older and younger rocks' thus revealing aspects of 'the geological history over the last 600 million years'. Not only qualitatively the 'best in the world', the cove rocks tell a story of South Australia spanning the top eighth of Earth history.[18] In support of Hall's contention, it is noteworthy that South Australia's only World Heritage listed place is the Naracoorte Caves, half of the Australian Fossil Mammals serial property, which is recognised for its superb late Neogene and Quaternary vertebrate fossil record. The National Heritage listed Ediacara Fossil Site is another celebrated South Australian deep time landscape, conjuring Precambrian sea floors from arid quartzite hillsides in the ancient Flinders Ranges.

The initial push to 'save' Hallett Cove came from academic geologists, some of whom were also members of the fledgling National Trust of South Australia, incensed at the proposed destruction of a canonical scientific site (a 'Mecca for visiting geologists' and 'the pride of local geologists'), and concerned to protect their outdoor laboratory to help train future geologists; and from South Australian heritage and school groups eager to preserve an outdoor museum of global significance. It was certainly an idea whose time had come: already in 1965, deserts were regarded as 'landform museums rather than scenes of rapid destruction' and Australia had been described as 'the world's foremost soil museum'. The first mention of the need to preserve the cove as a 'practical teaching area' might have occurred in December 1961, when Norman Tindale, then Chairman of the Nature Preservation Committee of the National Trust, proposed it at a committee meeting. Dean Berry provided a succinct account of the value the Trust placed in the site as an educational resource in a letter to Maud McBriar of 1 March 1967: 'The whole area shows unique evidence of Permian glaciation and there are exposures of other rocks spanning a period of more than 600 000 000 years. This makes it virtually an open air demonstration room for students of geology at all stages.' The 'cameo' of Hallett Cove shows Earth-history-in-the-making, conveniently near Adelaide. It provides the possibility for a series of deep-time vignettes: as SASTA explained in a 1971 report to the state government calling for the acquisition of Hallett Cove as major district open space (as proposed by the government's own Metropolitan Development Plan of 1967), 'there is more to teaching Earth Science than a simple examination of rocks. That much of the exercise can be performed in the class room. It is only when students see rocks in relation to other rocks and to the overall topography of the area that they develop the kinds of attitudes to the subject which all good teachers are striving to achieve.' It is this relational aspect, through time as well as space, that made Hallett Cove suitable as an open-air museum.

Campaigners like McBriar, Graham Whitten (as chairman of the GSA (SA)), Brian Daily (Geology lecturer at the University of Adelaide), and members of SASTA argued the case for preserving the Hallett Cove landscape as a training resource from early in the 1965 campaign, in part to establish their credentials among industry geologists and a public which was riding high on a national minerals boom. SASTA, in its Open Space report, explained that no site 'enables a student to take in such a wide range of Geological history in one excursion' as Hallett Cove, which is 'unique as a teaching aid', in that 'the same area can be given a simple interpretation suitable for primary school students or a complex interpretation posing problems at tertiary level'. Furthermore, it is 'still close to its natural condition and within a few miles of a capital city'. *Advertiser* columnist, writer and conservationist Harold Lindsay wrote in that newspaper in 1965 that in Hallett Cove, the state had 'valuable areas for the field studies which are essential for the training of geologists – and it is our geologists whose work leads to the discovery of mineral wealth'. An article in the *Advertiser* in late December 1967 explained that the cove had 'made possible extensive knowledge of Australian stratigraphy which had contributed, and would continue to contribute, to exploration for oil, gas and other minerals'. A letter to the *Australian* newspaper in September 1971 repeatedly stressed the (unspecified) '"mineral phenomena" of the cove that would be threatened by subdivision'.

This was not just political opportunism, as argued by the developers. On the one hand, it was necessary not to alienate those members of the Geological Society who relied on extractive industries for a livelihood. The distinction between industry and academic geologists is often blurred. Many scientists were active in the Geological Society, tireless campaigners for Hallett Cove, and also worked for a mining company. Others felt that the society should not oppose development as to do so would set a precedent which could have ramifications for the extractive industries. On the other hand, most campaigners were infused with genuine pedagogical commitment. Their belief in and commitment to the importance of the site as a teaching resource, as a monument to the history of the discipline, as a testimony to South Australia's deep past and as intrinsically valuable as a site of outstanding geological interest and global importance in its own right were unequivocal. However, a pragmatic public focus on the economic relevance of the scientific value of the site continued throughout the years of the campaign and into the 1980s.

Early recognition of its glacial relics brought the site to scientific notice in South Australia and kept it there during 100 years of controversy and exegesis. Its glaciological significance was central to the National Trust's efforts to secure the Sandison Reserve and remained important throughout subsequent campaigns. Those landforms still brand it as a 'premier', 'priceless', 'unequalled' or 'world class' locality for the study of the Permian Ice Age. But another string to its geological bow was increasingly articulated from 1968 and continues to be celebrated in recent cove literature – the significance of Hallett Cove as a piece of the four-dimensional mobile global jigsaw of plate tectonic theory. The glacial features, as described by SASTA:

> when considered along with similar evidence from the other continents of the Southern Hemisphere and peninsular India, provide strong confirmation of how the present continents packed together before the breakup of the primeval continent of Gondwanaland ... [I]mportant though the quality of these glaciated pavements is, it is the clear way in which they are displayed in relation to rocks both older and younger which makes Hallett Cove unique.

Further, notes from informal surveys by open day guides in 1976 and 1977 emphasise the degree of public interest in continental drift theory (for example, Figure 15).

In a feature in the *News* of 20 September 1971, Maud McBriar emphasised Hallett Cove's Gondwanan credentials for the elucidation of 'a lively theory': 'One exciting point is that these rocks relate this part of the world to continental drift – a lively theory that Australia was once

Figure 15: Gondwana logo, Friends of Hallett Cove brochure. *(Photograph K. Douglas)*

part of an ancient continent that embraced South Africa, India and Antarctica.' In its 1971 Open Space report SASTA submitted one of its strongest arguments for the protection of the cove 'as a well protected unit to enable intercontinental correlation', explaining its relevance to plate tectonic theory as a 'key in the development of theories of the Earth's changes as a whole and not just in a specific locality, state or continent'. An *Advertiser* article of 4 April 1975 quoted Minister Broomhill, speaking in support of the Save Hallett Cove Committee, which had presented to him 'new evidence' showing 'the importance of continental drift patterns at Hallett Cove'. It is worth noting that early in 1971 Broomhill had supported in principle an application to quarry the south end of the Amphitheatre. The minister's new-found fervour for the preservation of the cove might have been influenced by the letter which accompanied the Save Hallett Cove deputation, from Tom Uren, who had already pledged his support, and by the clear swing of public and political opinion in favour of preservation and protection.

A letter to the editor in the *Advertiser* of 22 September 1971 from R.F. Harris, a University of Adelaide Honours student in geology, pointed out that 'the 50-acre [20 hectare] site of scientific interest does not include all the important items at Hallett Cove. The cove, the foreshore and the adjacent cliffs have not been fully studied. These contain rock features not found elsewhere and the only fossil locality in glacial rocks at Hallett Cove. With the ... present plan all these will be bulldozed.' On 18 June of the same year, Harris sent a submission to Broomhill and the Director of Planning, Hart, setting out his case more comprehensively for the protection of 'certain unpreserved relics of our geological heritage' which have 'not been studied in the light of modern developments in sedimentary Geology'. Furthermore, 'much of the Permian rocks are imperfectly described, despite this area being "classic ground in Australian Geology". Thus the fact that areas have not been previously reserved does not mean such areas are unimportant.' This suggests a degree of disingenuity on Broomhill's part, as the letter from Harris demonstrates his awareness of the 'new evidence' four years before he moved to support the Save Hallett Cove Committee's demands for extending the park's boundaries.

Notes prepared for Uren on 10 September 1974 describe Hallett Cove, 'a national heritage', as 'a vital link' to other continents of the Southern Hemisphere 'in theories regarding the break-up of the primeval continent of Gondwanaland by the drifting apart of the present continental masses'. A *Sunday Mail* article announcing one of several public open days at the cove, on 17 October 1976, promised that 'geologists will show the secrets Hallett Cove has yielded about the age when the land mass that is now Australia broke away from Antarctica 45 million years ago and 40 million years later was submerged beneath the sea in this area'. In a submission from the Save Hallett Cove Committee and the Conservation Council of South Australia (representing 24 organisations with state-wide membership including the Nature Conservation Council, the National Trust and the Field Geology Club), Uren was told in no uncertain terms that Hallett Cove 'provides one of the finest pieces of geological evidence that this land was linked with Antarctica'. The submission was signed by as diverse a range of organisations as the Geological Society, the Hallett Cove Beach Progress Association, SASTA and the Gem and Mineral Society. This submission enticed with its potential for economic geology as well, with the claim that 'the theory of plate tectonics, involving continental drift, explains the present shape and distribution of continents on our earth and gives new insights into the search for mineral resources'. It is symptomatic of a philosophical change between the 1957–65 negotiations and the later campaigns, as explained in an advertisement by the Save Hallett Cove Committee in the *Advertiser* on 31 March 1975: 'When this was a rural district it seemed sufficient to hold a token strip of land along the cliff-top ... however the emphasis has changed to a more comprehensive study of the geology here ... lands which lie outside the present buffer zone will continue to yield more scientific data as they are further researched for their geological and archaeological contents.'

Both academic and popular literatures – the latter in the form of Friends of the South Australian Museum (FOSAM) and NPWS brochures – privilege the site's geological importance over other aspects of its history and heritage. McBriar wrote candidly in 1971 that 'The case for reservation of an additional area of land was based almost solely on the need to preserve the geological features, although some consideration was given to protecting the relics of the former aboriginal inhabitation.' At the same time, a SASTA submission complained that 'the [flora] and the anthropological importance of the cove are too easily overlooked'. As described earlier, humanity is sometimes written out of cove history altogether, in an eagerness to play up its wilderness credentials: 'the loneliness of the deserted beach, the long stretches of nature undefiled and timeless'. But in contrast to the Little Desert National Park, the need to preserve aspects of the human past of Hallett Cove has also been integral both to the campaign and to subsequent interpretation of the site. For example, the Worthing Mine site is recognised as a relic of mineral exploitation in South Australia. Graeme Pretty, then the President of the Anthropological Society of South Australia, wrote to McBriar on 6 May 1975, citing the importance of the archaeological material above Waterfall Creek as a site of 'Ancient Man at Hallett Cove, notwithstanding [its] lack of ruined walls and street alignments'. He envisaged that, properly protected, Hallett Cove 'will enable Australia to memorialise Ancient Man's way of life in a unique way'. As described by Robin of the Little Desert campaign, the human heritage (both Indigenous and European) that is venerated at the cove is a humanity of the past, not a celebration of living culture. It is a form of cultural value-adding to natural heritage. As Ron Caldicott expressed it in 1974, 'The presence of relics of former Aboriginal inhabitation and other settlement also add to the value of the Cove.'

The campaign was unprecedented in the history of the Geological Society of South Australia. McBriar and other pioneers in the field of geological heritage in Australia borrowed many of their ideas from the Nature Conservancy Council during several visits to the UK in the 1960s.[19] The Geological Monuments Subcommittee of the Geological Society of SA was formed largely as a result of the Society's involvement with the Save Hallett Cove Campaign, and the battle marked an early fight for the newly formed Nature Preservation Committee of the National Trust of South Australia. The subcommittee, subsequently re-named Geological Heritage, first met on 10 March 1967, and consisted of Arthur Alderman, Brian Daily, Graham Whitten, and geologist Peter Allchurch. Maud McBriar was convenor. McBriar and Whitten were also representatives of the National Trust's Hallett Cove Subcommittee from 1966 to 1971.[20]

There are now Standing Committees on Geological Heritage in every state division of the GSA. Professional geologists are not traditionally considered to be pro-conservation, but in this fight, academia, government and industry banded together to protect both the history of the discipline and the landscape history embedded in the cove. There are many within the Society's ranks who have opposed heritage listing, on the grounds that it is not in the best interest of the mining industry, as McBriar, in her role as Convenor of the Geological Monuments Subcommittee, explained to Hart, the Director of Planning, in July 1971:

> Geologists within the Society are divided on the proposal being canvassed by various conservation bodies and the Conservation Council of South Australia for the creation of a Major District Open Space in the area. The Society has not adopted any policy on this matter but has left it to individual members … many have signed the petition that is currently being circulated, but others do not support it.

Furthermore, some Earth scientists are concerned that the different demands of geology and protected areas may impose restrictions on research. The Hallett Cove Conservation Park and the Sandison Reserve were primarily preserved for the cove's value as a geological site, but the park is managed by the National Parks and Wildlife Service – which is opposed to extractive industries on reserved land. In this case geologists and NPWS staff seem to handle conflict arising from this potential inconsistency. Traditional forms of protected area management are often considered inappropriate for the needs of a geological site and tensions abound where the needs of researchers clash with the anti-extraction policies of National Parks management.[21] And for all its efforts in the name of geological conservation, the Geological Heritage Subcommittee of the GSA (SA) has only once directly opposed a mining interest in the state – sand-mining of the dunes along Yankalilla Bay in the vicinity of Normanville.[22]

Hallett Cove is a celebrated but equivocal success. Its proximity to densely settled suburbs is a mixed blessing. Because of the significance of allochthonous rocks, all the walkways in the park are wooden, to avoid contamination. On-site and online interpretation of the glacial, plate tectonic and intellectual history of the cove to the public is ongoing, and integrated with recognition of Aboriginal history and ecology at the site. It still functions as an 'open-air museum'. But some campaigners maintain that the scientific value of the park and reserve is compromised as potentially important sites whose geology has not been investigated were and continue to be lost to housing. Vandalism has been a problem at times, particularly the damaging of interpretive signs and dioramas. There is a conflict between the management requirements of a conservation park, a scientific resource, and the need for urban public spaces. Given the density of the settlement around Hallett Cove, its custodians fear that recreational use will spill over into the geological reserve and put extra pressures on the site, not to mention the potential ravages of unrestricted amateur or illicit commercial rock-collecting. In 1976, the Rudman Report recommended the creation of a public reserve a little distance from the conservation park to avoid such overlap, but this did not eventuate. However, the South Australian Department of Environment and Heritage's National Parks code of conduct specifies that visitors should respect geological sites.

Hallett Cove and geological knowledge

thus has the whirligig of time brought in revenges to the old prophets[1]

Edgeworth David's precocious epitaph for the glaciation controversy in Australia fittingly introduces a concluding reflection to this essay. Hallett Cove and glaciation form the connecting threads in an investigation of the power of deep time to rescue and canonise a landscape. What does this story tell about the way the Earth sciences are made? Geological knowledge, like all knowledge, is made in part by processes of negotiation and consensus, and powerfully informed by personal experience. Similarly, the aesthetic and heritage value of a place is a matter of public negotiation and consent.

The contribution of the material landscape of Hallett Cove to debates about time depth in Australia is partly an artefact of the dependence of the Earth sciences on the physical remains of the past. The more information physically available to geologists, the more that can be deduced about the material Earth. Increasing professionalisation of the Earth sciences in Australia, technological innovation (particularly the dating revolution brought about by the study of radioactive isotopes) and associated change in intellectual agendas also contributed to the production of geological knowledge at Hallett Cove during 120 years of investigation.

As more information was obtained from and about the world, so were more sophisticated and precise conjectures possible. The sheer quantity of information necessitated a greater degree of specialisation within the discipline. Ralph Tate, Walter Howchin and Edgeworth David were geological jacks-of-all-trades with eclectic sub-disciplinary proficiencies that are luxuries rarely available to modern geologists. This is not to imply a deterministic progression from state of ignorance to one of enlightenment, simply that in a field science such as geology, the discovery of new territory, new material remains of the past and the development of new technology contribute to a fuller, though not necessarily more accurate, understanding of it.

Informed by a geological aesthetic and by the history of their discipline, Hallett Cove has been a significant site for South Australian Earth scientists almost since Tate's discovery of grooved or striated glacial pavements at Black Cliff, both for reasons of intellectual genealogy and heritage and for its physical properties. Because Hallett Cove was regarded as a canonical geological place in South Australia and because of its apparent isolation, it was not immediately obvious to most geologists that the site would need protection from the encroachment of urban sprawl, although as Harold Lindsay noted in 1965, 'on numerous occasions, the late Sir Douglas Mawson … and other geologists advocated that the Government should take steps to preserve the glacial pavement. Botanists added their pleas, stating that on the area some interesting native plants were surviving which appeared to be extinct elsewhere. Nothing was done however.' On the other hand, developers failed to appreciate beauty that could not be improved by sensitive urban planning, subdivision and landscaping in such an expanse of soaring land values. The 17-year campaign to preserve the geological integrity of Hallett Cove may have helped to effect an appreciation and awareness of geology, deep time and natural heritage in South Australians. But contrary to the sometimes triumphalist rhetoric of the anti-development campaigners, it is not intrinsic value that made the cove newsworthy or salvageable. Rather, a complex interplay of events influenced the course of the campaign to help the

Geological Society and the National Trust secure popular support to save a landscape that in the eyes of many could have been any other.

South Australian parochialism and the trope of uniqueness helped mobilise public opinion against urban development at Hallett Cove. At least two of the companies planning to subdivide the cove during the 1968–1975 period were Western Australian (Kadima Pty Ltd and Silesia Pty Ltd). This was belaboured in the media. Public and media demands for accountability, the buzz-phrases 'ecology', 'national estate' and 'urban sprawl', the rapid advocacy of the cove by establishment bodies with clout during the early stages of the campaign, the effective mobilisation of the romance of deep time in the reconstruction of Hallett Cove's interlocking constellation of theme park landscapes, all these contributed to the eventual reservation of 50 hectares of cove land in 1976. Whether campaigners tapped into burgeoning community interest in conservation or the *Zeitgeist* caught up, the resonance between the deep past as represented by its material remains at Hallett Cove and the concept of natural heritage as part of a National Estate captured public, media and eventually bureaucratic imaginations, and continues to play an important part in the articulation of philosophies of geological heritage, conservation and site management in South Australia.

The intrusion of Hallett Cove's municipal history into its intellectual and geological stories provided opportunity for Earth scientists and heritage professionals to mobilise the deep past, its material remains and its intellectual legacy, in the interests of geoheritage and the establishment of a collective national treasure chest of places worth saving. Hallett Cove as a storehouse of information about Australia's deep past is thus a national resource. Those involved in the final stage of the conservation battle described in the previous chapter tended to represent the need for an expanded park in terms of the earlier failure on the part of geologists, the National Trust, government and the public to understand the heritage importance of the unreserved areas, which they all now appreciate. It is not so much that faulty comprehension was illuminated, rather that value in the landscape was increasingly re-inscribed in geological terms during the late 1960s and early 1970s. It simply became easier for the geological community to alert the South Australian public to the importance of the cove landscapes.

The aggregate that makes up Hallett Cove reflects more than 500 million years of landscape evolution. Its outdoor museum is a shifting stage in an ongoing process of deposition, uplift and erosion. An overburden of cultural and social history is apparent in stone tools on the cliff tops, Aboriginal landscape histories, remnants of beach huts, floral distribution, wooden walkways and fences, and the housing estates which now press at the hems of the conservation park. Its decoding has a genealogy reflecting intellectual trends, patronage, technological innovation and switching allegiances, both personal and national.

The story of deep time at the cove is broader than debates about glaciation or controversy about zoning. For geologists, it is the clear conjunction of former worlds and the global importance of its Gondwanan inheritance that are drawcards as much as the specifically glacial features preserved within the park's boundaries. Its theme park nature is unusual in Australia. The area records the legacy of mountains and seas, sites of Aboriginal and European settlement and encounters, smuggling, copper mining, subdivisions, conservationists, beach huts, botanists, geology students and an oil refinery at Point Stanvac. Such a landscape is a composite of chemical, mechanical and social legacies, formed and conceived in different ways at different times. But to geologists viewing landforms in the park, the clear association of several lost geographies takes on a global significance as the slow dance of former continents becomes the bedrock of time to come.

Part Two

Dirt, bones and the Diprotodons of Lake Callabonna: discovering the lost worlds of vertebrate palaeontology

They are finding bits of the jigsaw, putting the pieces back together, peering at a lost world through an ever-widening window

Elizabeth Smith, 2001

1892

During October 1892, welcome rain soaked Adelaide and its arable hinterland. Appreciation for 'the grand season' flooded in to the *Adelaide Observer.* At Innaminka on 14 October, correspondents celebrated the Farina mail, 22 hours late because of the splendid rain. After half an inch fell in the colony's capital in the week to 3 November, farmers in Yardea on the northern Eyre Peninsula were even 'anxious to see fine weather now, so as to enable them to get in their hay which is fit to cut'. From Blinman it was reported that all dams were full and all creeks running; 'a very unusual thing' in October. To drought-blighted pastoralists on the Eyre Peninsula, the growth of feed was 'most phenomenal' and in Orroroo, 'such growth of feed and crops' had 'never been seen before'. The North was saved for the season.

On Tuesday 1 November across the border in Victoria, crowds of visiting South Australians watched rank outsider Glenloth win the Melbourne Cup on a heavy track, and the fashion pages that week reported a daring craze for plaid. The same afternoon in Adelaide Edward Charles Stirling, in his capacity as the Director of the Adelaide Museum, delivered a well-attended paper to the Royal Society of South Australia discussing a commemorative copper tablet which Matthew Flinders left at Cape Catastrophe. The museum had recently found it in a drawer. He little realised that by the end of the week his attention, and the city's, would turn to found objects buried far deeper in South Australia's history and of possibly greater significance for its future.

Adelaide had spring in her step that Tuesday afternoon as John Meldrum rode through town after some 500 miles (800 kilometres) hard travel, from the Ragless brothers' Callabonna Station on the eastern shore of Lake Mulligan, near the colony's border with northern New South Wales. Cutting a path to the city through desert turned to fen, through a wheat belt suffering the attentions of too little rain at the wrong time, along the muddy main road from Gawler, the well-digger came burdened with material suggesting that such unexpected weather as impeded his passage might once have been commonplace in the arid north-east of the colony. His discovery of the fossil remains of gigantic animals allowed the possibility that conditions in former times were different from the climate of modern South Australia.

These implications might have escaped Meldrum, but he knew that the news he had to relate and the specimens he carried with him, including the lower jaw of the giant extinct marsupial Diprotodon 'with some of the molars attached', would attract the attention of 'scientific men' in Adelaide. He placed them on view at Mr Lee's Black Swan Hotel on North Terrace (Figure 16). The bones had suffered on the long trek to Adelaide and so had limited intrinsic value as collectibles. Their importance lay rather in 'the proof of the existence of the giant, extinct marsupial of centuries ago in a locality in which its fossil remains have not hitherto been discovered', and in the tantalising implications of this 'proof' that the desert had once supported large marsupial herbivores in great quantity.[1]

Such images or the promise of lost worlds pervade the language of palaeontology. The material remains of the deep past are catalysts to the imagination. The motif of a lost or former world of fantastic or ferocious beasts now gone is rehearsed and recycled in various formats, elegiac, educational or sensational. It is expressed in academic and popular science, museum exhibits, film, television and art. Fossils and palaeoclimatic data are pieces of a jigsaw, which the reconstruction, diorama or scene makes intelligible. Fossils and other material remains of the past 'tell'; they evoke, echo and stimulate the palaeontological imagination to build scenes,

Figure 16: North Terrace, Adelaide, looking towards the Institute and University. *(Samuel White Sweet, between 1869 and 1889. National Library of Australia nla.pic–an20886593–51)*

vistas, reconstruct whole 'worlds'. The small cog provided by John Meldrum meshed later with other evidence from the lake, which helped scientists articulate 'the theory ... that in some far back period there was a vast lacustrine region in the locality named, and the diprotodon and his kind were attracted to this green spot and died from want of food'.[2]

The marsupial 'mammoths' of Australian prehistory, Diprotodons were first noticed by Europeans in 1829 in the fossiliferous Wellington Caves. They continued to intrigue naturalists at the end of the century, as their foot, tail and epipubic bones remained elusive, obscuring their habits and nearest living relatives. The institution or individual that first obtained a complete skeleton, unknown in over 60 years, could expect considerable prestige and financial reward. A persistent rumour among geologists and palaeontologists who have worked at the site, and South Australian Museum workers, suggests that either the board of the South Australian Museum or the Royal Society of South Australia promised a £1000 reward to the first person who produced either a complete Diprotodon skeleton, or the mysterious feet, but I have been unable to find a contemporary report. It does not seem likely that a cash-strapped public institution during the 1880s and 1890s could afford such a reward, even given the windfall promised by a complete Diprotodon.[3]

The specimens at the Black Swan Hotel and Meldrum's description of other, more complete skeletons on the lake surface away beyond the ranges provided a wedge, not just into the planet's history, but into bureaucratic purses, which understood that the promise offered by the bones could be realised with an expensive, administratively and strategically complex expedition to the remote north. Adelaide was a city-state at the heart of a great wilderness as the nineteenth century turned into the twentieth. Adelaidians, self-consciously liberal, comfortably middle-class and mildly philanthropic, believed in education, having founded both a museum (in 1856, under naturalist curator George Waterhouse) and Australia's third university (in 1874) before they sewered the town. Encouraged by the world opened up by Meldrum's Diprotodon bones, by the promise of wealth, of knowledge and of kudos for the colony's cultural institutions, articles in the *Observer* and the *Register* exhorted the museum board to send an excavation team to investigate Lake Mulligan as soon as possible. The newspapers urged that Edward Stirling and the museum's assistant director A.H.C. (Amandus) Zietz 'be laid on their track' with the utmost haste.[4]

Part Two describes the discovery and articulation of lost and imagined worlds in Australia's geologically recent past, 280 million years after the glaciers retreated from Hallett Cove leaving scratched bedrock, sediments and foreign boulders in their wake. It begins with the public introduction of the Lake Mulligan–Callabonna Station Diprotodon remains in Adelaide in 1892. The lost world is the contemporary geographic or scientific frontier waiting for discovery and enlightenment to turn it into a scientific site, as occurred at Lake Mulligan in 1892–94. The site itself may then disappear from scientific consciousness to exemplify a different type of lost world. In this sense, the lake has been lost and found many times during its century-long intersection with the history of Australian palaeontology.

Chapter 4

THE LOST WORLDS OF LAKE CALLABONNA, 160 000 000 BP – 1970

Losing Lake Mulligan

On 30 November 1901, Hallam, Lord Tennyson, first Governor of the newly federated South Australia, reserved Lake Callabonna, formerly Lake Mulligan, a dry 'lake' south of Lake Eyre on South Australia's arid eastern boundary 'hereto for the purpose of the preservation of Fossils'. The *Government Gazette* of 5 December proclaimed its boundaries by virtue of the provisions of the *Crown Lands Act, 1888* (see Figure 17).[1] Feted for nearly a decade for its rich

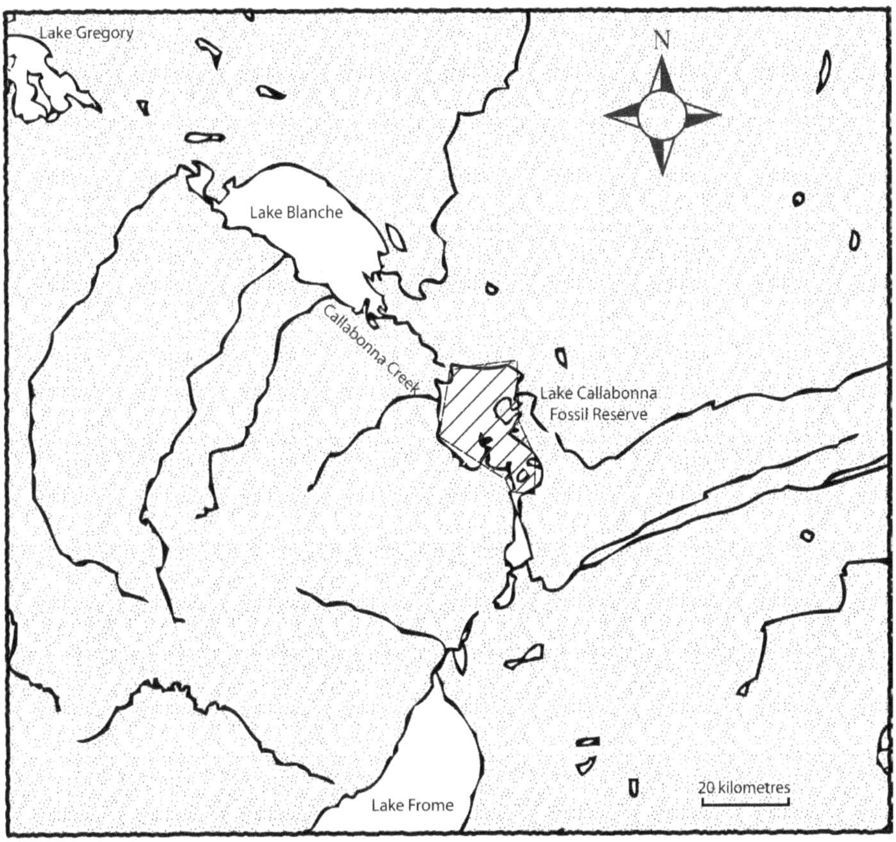

Figure 17: Lake Callabonna Fossil Reserve. *(Image Brett Calcott)*

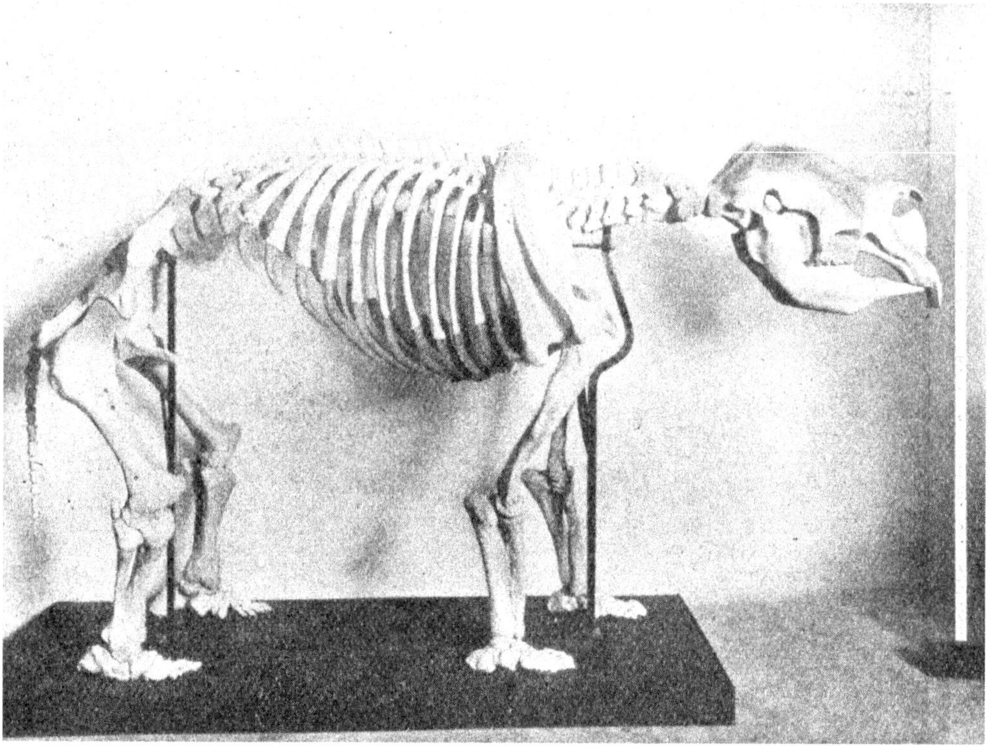

Figure 18: Plaster cast of the skeleton of *Diprotodon australis*, South Australian Museum. *(Reconstruction of* Diprotodon optatum, *from the Callabonna Deposits, South Australia. Reprinted by permission from Macmillan Publishers Ltd: Figure 1. In E.C. Stirling, 1907,* Nature, *76(1978): p. 543, copyright 1907. National Library of Australia S 505 NAT (1544216))*

burden of giant, extinct marsupial and bird fossils and the palaeoecological surmises this encouraged, nonetheless the lake had not seen a scientific expedition since the South Australian Museum packed up its boxes, bones and camels in 1895. Assembly of its famously 'complete' Diprotodon reconstruction – the expedition's object and prize – was not finished for another six years.

E.C. Stirling, in collaboration with Amandus Zietz, studied and assembled this skeleton of the rhinoceros-sized browser, the largest species of pouched mammal unearthed in 170 years of Australian vertebrate palaeontology, from a miscellany of Lake Callabonna remains brought to the museum between 1892 and 1895. This composite skeleton was never displayed, as the salt-damaged bones were too fragile to hold their own weight. Instead, the museum preparator Robert Limb performed the Herculean task of reproducing 'every significant bone in plaster', then hand-coloured each to resemble bone. The Museum unveiled the cast in 1907, the year after its jubilee celebrations, and the Diprotodon skeleton graced the foyer for many years (see Figure 18).[2]

Lake Callabonna is one of a chain of large saltpans between Lake Eyre to the north-west and the Flinders Ranges to the south. Stranded beaches and fossil fish, turtles, waterbirds and bivalve shells indicate that formerly these lakes were much larger and intermittently filled with fresh water. They arc in a horseshoe from Lake Torrens in the west to Lake Frome in the south-east. Edward John Eyre, who travelled in the region in the late 1830s and 1840 hoping to find a route to the centre of Australia, assumed the discontinuous horseshoe was a single giant lake, formerly an arm of Spencer Gulf, which he named Lake Torrens (see Figure 19). On 2 September 1840, looking out from the hill he had pessimistically named Mt Hopeless, a 'low haycock-like peak' to the north-east of the Flinders Ranges, he 'at last ascertained beyond all doubt' that the basin of 'Lake Torrens' commenced near the head of Spencer Gulf, bending around north of the Flinders Ranges and southward, almost isolating 'the island of South Australia, for such I imagine it once to have been'.

From 1840 until roughly 1860, the horseshoe remained unbroken in settler imagination. Charles Sturt's Central Australian expedition passed close to the channel between Lakes Callabonna and Blanche more than once in 1844 and 1845 but the explorer, at first hopefully and then in frustration, reaffirmed Eyre's identification of the large depressions as the eastern arm of a single enormous lake. He explained in his field journal that 'the desolate barrenness, the dreary monotony, the denuded aspect of this spot is beyonde description' (see Figure 20). A number of other surveyors and explorers to follow Eyre rehearsed his mistaken identification, until B.H. Babbage and A.C. Gregory in two separate expeditions in 1858 began the process of fracturing Lake Torrens into its component saltpans, and on their heels came settlers. From 1860 until 1894 South Australians knew Lake Callabonna as Mulligan Springs, Lake Mulligan or Mullachon after an Aboriginal word for the mound springs on its western side, and it was distinguished only by its position close to the northern limit of the pastoral frontier. An Aboriginal stockman, possibly called Jackie Nolan, might have indirectly alerted the Adelaide scientific establishment to its palaeontological riches in 1892 by showing them to the Callabonna Station leaseholder Fred Ragless, or to John Meldrum, who then brought samples to Adelaide.[3]

In mid-1893 E.C. Stirling moved to rename the lake Callabonna after the creek which sometimes flows into it from the Strzelecki, and from which the sheep station took its name. He mistook it for an Indigenous word and regarded it as 'more euphonious' than the Irish-sounding Mulligan. His offering contributed to a public debate in the Adelaide media about renaming Lake Mulligan, which saw suggestions ranging from Lake Stirling in recognition of the museum director's celebrity, to the light-hearted claim that shedding Mulligan was a pre-emptive move on the part of English settlers to write the Irish out of South Australian history.

In colonial Australia, fossils were part of the capital of the earth. Callabonna's early post-excavation history describes an emerging scientific independence in Australia. The importance of material artefacts in the reconstruction of the past demanded a conversation between metropolitan and colonial scientists (Stirling and Zietz), their informants ('Jackie Nolan', Meldrum and the Raglesses) and the dirt itself. This hierarchy of negotiation eventually undermined the authority of the distant metropolis. The roles of patron and client were readjusted. Rationalistic centre–periphery models characterise the colonial knowledge trade as radial and recursive: 'raw' objects and data were passed from wandering colonial informants and naturalists to metropolitan savants and scientists, who transformed them into knowledge and theory which they exported back to the periphery, and so on, for example, as in George Basalla's seminal 1982 article published in *Science,* 'The spread of western science'. Such models often overlook and elide the presence or emergence of other centres and other peripheries within and around 'the' metropolis and colonies, neglect the importance of cross-cultural encounters and exchanges, ignore the nuances of personal, national, intellectual and institutional loyalties and agendas, and forget the environmental particularity of places.[4]

Adelaide itself, the undisputed political, intellectual, commercial, administrative and ecclesiastic heart of South Australia, was both metropolis and colony. Furthermore, many 'colonial' naturalists had powerful metropolitan patrons and institutional attachments outside

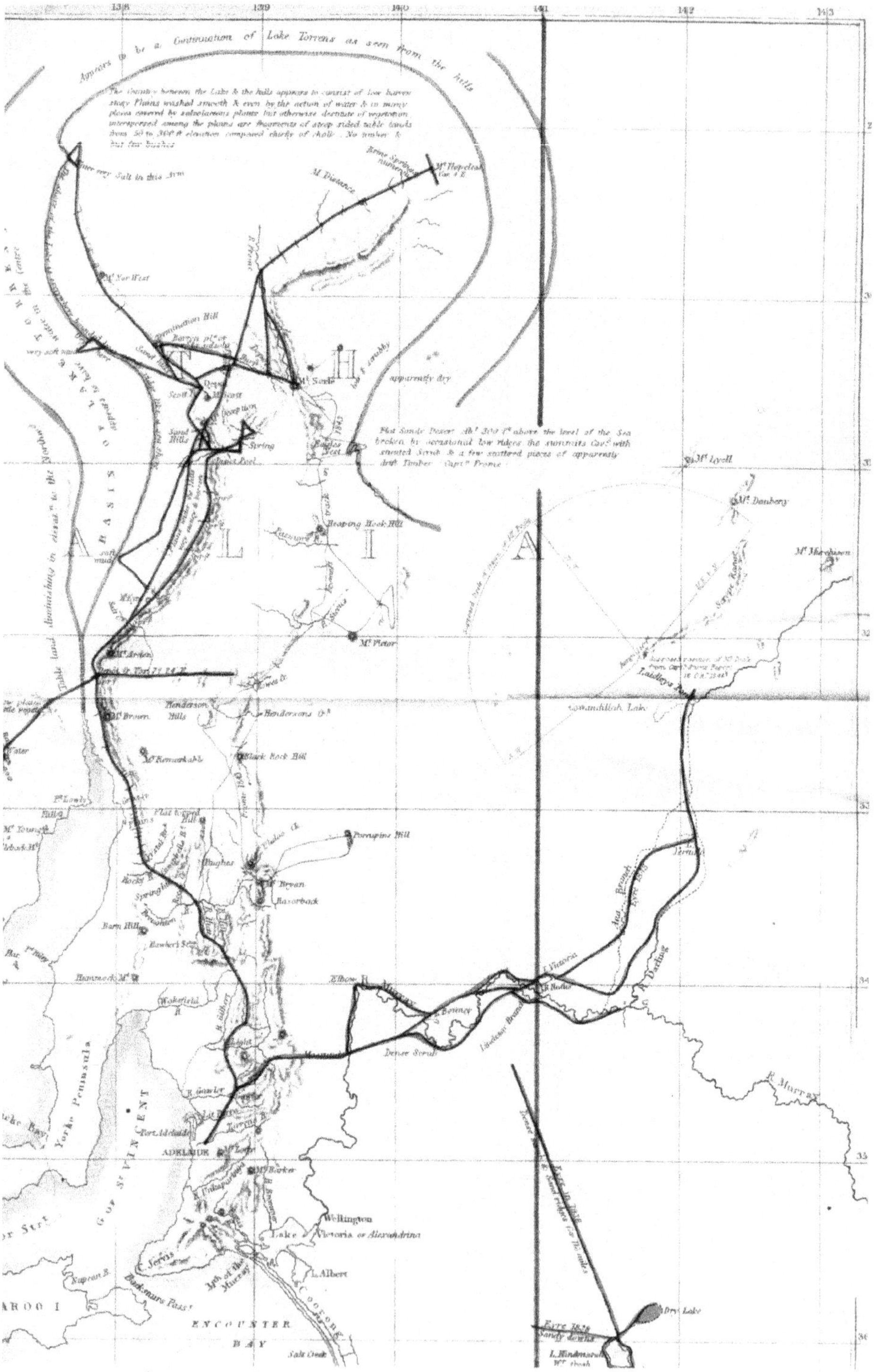

Figure 19: E.J. Eyre's horseshoe lake. *(Extract from John Arrowsmith, 1845. Map of Mr Eyre's routes into Central Australia, and overland from Adelaide to King George's Sound, 1840–1841)*

Figure 20: Charles Sturt's party views the eastern shore of 'Lake Torrens', 1845. This is probably the southern shore of Lake Blanche or Lake Callabonna. (The sandy ridges of Central Australia. *Samuel Thomas Gill, 1846. National Library of Australia nla.pic–an2377285)*

the British intellectual centres of London, Edinburgh and Oxbridge, and were also themselves committed to fostering scientific institutions and societies in the Australian colonies. Edward Stirling, for instance, attended Trinity College, Cambridge, from 1865 to 1870, was a member of the Royal College of Surgeons from 1872, elected Fellow in 1877, and received a DSc from Cambridge in 1910. But he was born in Adelaide and educated there at Saint Peter's College. He returned to South Australia in 1881 after years in the United Kingdom, Iceland and Melbourne, and from 1884 to 1887 he represented North Adelaide in the Legislative Council. He founded the University of Adelaide Medical School. He was Director of the South Australian Museum from 1889 to 1913 and President of the Royal Society of South Australia in 1889. His colonial loyalties, particularly to those institutions he patronised, provided a powerful check or complement to his often quite disparate metropolitan institutional affinities.

Callabonna's geographical inaccessibility and the problems associated with excavation and preservation of the fossils (which still have not been entirely solved) offset to some extent the wealth of its 'veritable necropolis' of giant marsupial bones.[5] A degree of scientific neglect resulted and the physical site has slipped in and out of the palaeontological canon over the last century, subject to the vagaries of climate, economic policy and intellectual fashion: Berkeley palaeontologist Richard Tedford observed in 1973, 'The richness of the site was apparent, its palaeontological importance was unquestionable, and its geological significance had been hinted at, but no further studies were conducted there.'

Notions of deep time are conditioned by intellectual, economic, cultural and geographical circumstance. Compared to Lake Callabonna, and notwithstanding comparable difficulties of access, the Cenozoic fossil localities of the Lake Eyre basin underwent something of a renaissance from the 1950s–1970s, as their place in the global jigsaw of palaeontological knowledge was conceived, investigated and celebrated.[6] The importance of the Callabonna site in the 1890s and again in 1953 and in the 1970s hinged on contemporary international, intellectual and climatic contingency: scientific narrative intersects with social and environmental history. Highlighting the serendipity of palaeontological discovery, every expedition to Callabonna that has returned with a significant bounty of fossils was undertaken at the tail end of several years of drought, before the site turned into a rainy quagmire impassable to truck and camel alike, in eerie imitation of the entrapment scenario evoked to explain the presence of so many prehistoric bones.

Scientific work at Lake Callabonna resumed in 1953 when the South Australian Museum persuaded Professor Ruben Stirton, Director of the University of California Museum of Paleontology (UCMP) at Berkeley, to detour to the lake in the course of his search for much older mammalian fossils in Australia. Stirton had received Fulbright funding for himself and a student, Dick Tedford, to investigate possible fossil affinities in Australian Tertiary sediments with the Colombian fauna that had so fascinated him throughout the 1940s. The South Australian Museum agreed to provide vehicles and a preparator, Paul Lawson, if the Berkeley expedition diverted to Lake Callabonna to investigate the status of the site – whether the fossils had eroded away or been silted over as local knowledge insisted – and to establish whether the 1901 boundaries of the Fossil Reserve actually coincided with the location of Stirling's 'necropolis', since 'an aura of mystery began to surround the original sites':[7]

> Local property-owners, including the Ragless family [which had held the Callabonna Station lease in the 1890s], believed that the bones once exposed on the lake floor had been covered by silt and were no longer visible. In 1953 our hosts, members of the South Australian museum board, agreed to help us in our search if we would visit Lake Callabonna and try to locate the original excavation sites.[8]

More importantly for the lake's place in the palaeontological canon, the experience had a profound effect on the young Dick Tedford, who wrote his doctorate on another site visited on that first of many Australian expeditions, but remained fascinated by the hints Callabonna's evocative present offered about its rich past. In 1973 he wrote of his younger self that he quickly recognised 'that this unique site would richly repay further work'. In 2000, his enthusiasm for the site remained undimmed:

> The array of fossils in that deposit is really quite incredible and would yield absolutely incredible reconstructions of the palaeoecology, but it's never been done ... Stirling and Zietz collected a much more diverse fauna than we did in 1953 ... there's quite a diverse kangaroo fauna there ... We also were aware that Zietz had explored more widely on the lake and found material 'everywhere we went thither' he said! ... when the lake surface was dry enough they were able to explore rather widely ... So I thought ... in 1970, this was between major La Niñas so the lake floor was really dry and we thought, this is an opportunity ... That trip really opened my eyes to what was there at Lake Callabonna, and it was mind-boggling.[9]

After the 1970 expedition, the site fell into obscurity again, celebrated but rarely visited, an interesting palaeontological margin note awaiting elucidation, a persistent memory in the specimen drawers and display cabinets of the South Australian Museum.

What is there

In March country at the northern end of the Flinders Ranges is red and brown, black and grey and blue. It is windy and dirty and frequently dry. The vegetation is tenacious but ephemeral, like the signs of human occupation. Creek beds tell of other hydrological regimes when rivers flow and the dust turns to mire. Still older stories weather out of blowouts, dunes and the banks of the creeks. The land is saturated with bones: big bones, little bones, new bones, old bones, deep bones, shallow bones. Speaking of bones, the lake, which is still further north and closer to the nation's topographic navel, is not a real lake. It is the bones of a lake. It is not even good sheep country, it is the carcass of a pastoral industry that died of thirst, its corpse mauled by rabbits.

In 1945, an eyewitness, Harold Fletcher, Curator of Vertebrate Palaeontology at the Australian Museum, described a cabbalistic, provocative place, almost the repudiation of landscape:

> At times during the day we had glimpses of the dry lake bed of Lake Callabonna ... The dry lake surface ... stretched away ... as far as the eye could see. It was completely devoid of vegetation and produced a feeling of gloom and extreme loneliness ... Even bird life was practically non-existent, and this introduced a hushed stillness adding an aura of mystery to the desolate countryside ... Away in the distance, towards the centre of the lake, a cluster of vegetation was miraged up, seemingly to float in the air.[10]

This lake is overwritten by another, gloomy and mysterious, emerging from the author's understanding of an empty landscape and a deeper palaeontological past. But before it was known as a graveyard there were other worlds there: inland sea, horseshoe lake, saltbush plain, swamp, desert, too boggy, too dry, wasteland, midden, refuge, larder. Such stories intersect and inform the lake's scientific stories. Meaning is as unstable as the lake surface.

In 1860, Robert Stuckey, a South Australian pastoralist, reported his discovery of the mound springs at Lake Mulligan. Of course he was not the first to find them. The lake is saturated with human presence through time. Middens containing worked flakes and other stone artefacts, eggshell fragments and bird bones abound on and near the springs. Some of the eggshells have been carbon-dated to between 2000 and 3000 years old. In the 1980s archaeologist Mike Smith found evidence of occupation in northern Pirlatapa country, in the Strzelecki Desert to the north of the lake, dated to more than 14 000 years ago. Fletcher himself passed over many clay pans in the course of a day searching for Diprotodon remains. They 'brought to light discarded or lost aboriginal implements consisting of flaked or chipped knives, scrapers, spear heads and, now and again, stone axes', evidence from the deep past which for Fletcher proved 'that this area of the country, in the past, was inhabited by tribes of aborigines'. Stuckey reported that local Aboriginal people in 1868 called the area Mulligan or Mullachon after the springs. 'The past' turns out to be shallower than Fletcher might have expected.

The watered inland

There are signs of water everywhere you care to look in the interior, on maps, in atlases, in dry lakes and creek beds, in relictual stands of rain-loving flora, underground, in mound springs and aquifers, in bones and sediments. Geologists, botanists, zoologists, ecologists, ethnologists, folklorists and Aboriginal people tell many stories about the onset of aridity. Here is one of them.

The Eromanga or Great Australian artesian basin has modern surface expression in the Lake Eyre internal drainage system. Its geological history extends from the Jurassic period, about 160 million years ago, when the centre of the continent began to subside. Gondwana still comprised India, Africa, Madagascar, New Zealand and South America, as well as Australia and Antarctica, but the Mesozoic supercontinent had begun to separate into smaller landmasses. Shallow intracratonic basins fed by river and lake systems developed with this rifting process. Giant reptiles lived in rivers, oceans and lakes, in the air and dinosaurs lived on land. Conifer, cycad and palm forests and fern prairies blanketed the landscape. The first true eutherian (placental) and metatherian (marsupial) mammal remains appear in the fossil record during the succeeding Cretaceous period (about 135–65 million years ago).[11]

These Lilliputian Mesozoic mammals appear to have diversified and expanded into many of the niches formerly occupied by dinosaurs after the demise of the latter at the end of the Cretaceous. It is their larger Neogene inheritors that concern me (see Figure 2). Returning to late in Panel 5 (Chapter 1), as Antarctica and Australia had finally separated and flowering plants, mammals and birds diversified, Lake Callabonna was part of a huge intracontinental river basin.

Panel 5

During the late Palaeocene epoch in the absence as yet of a circumpolar current, warmer ocean waters and a more temperate climate with seasonally high rainfall affected Australia, despite its southern latitude. As a rift valley began to separate Australia from Antarctica, fluviatile sediments from extensive river systems accumulated within the saucer-like intracratonic depression known today as the Lake Eyre basin. Pollen and plant fossils suggest that closed canopy, cool temperate rainforest, akin to that which survives in Chile and Tasmania today, flanked these rivers, dominated by *Nothofagus* (southern beech), *Araucaria* and celery-top pine.[12]

The faunas of this period are very poorly known. The mammalian faunas of the succeeding Eocene epoch in Australia remained largely unknown as well until the early 1980s. Excavation by members of the Queensland and Australian Museums at Murgon in south-eastern Queensland revealed a thin band of marsupial teeth and bones laid down in what appears to have been a shallow lake basin estimated at 55 million years ago. These animals seem to be allied to South American groups, which suggests a New World origin for Australia's marsupials. Many turtle and crocodile bones, fish, snake, lizard and frog bones, the bones of the world's oldest known songbirds, one of the oldest bats (*Australonycteris clarkae*), and, controversially, bones with the characteristics of eutherians (*Tingamarra porterorum*) accompany the Eocene marsupial remains.

At the end of the Eocene, the zone of cool temperate forest stretched from southern South America, through Antarctica, to south-eastern Australia, but increasing aridity led to a reduction in fluvial activity in Australia as the circumpolar Southern Ocean opened. Many modern groups of Australian biota appeared by the succeeding Oligocene epoch. Over the next 40 million years, as Australia drifted north and parted company with Antarctica, the migration pathways for terrestrial animals terminated, but flowering plants, grasses, birds and mammals radiated and diversified. Late Oligocene and early Miocene sediments record the expansion of plants of the Myrtaceae family, including the ubiquitous *Eucalyptus* and *Melaleuca*. The replacement of closed canopy conifer and rainforest by eucalyptus, with its accompanying understorey of scleromorphic shrubs, indicates a dryer climate.

A magnesium-rich megalake occupied much of central Australia during the early Neogene period, remnants of which persist in the Lake Eyre basin today. Climate change and uplift along the Great Divide in the east and along the southern margin of the Lakes Eyre and Callabonna basins led to renewed sedimentation. The first record of abundant grass pollen dates from this period, suggesting grasslands, which could indicate a drying climate. A varied fossil assemblage of diprotodont marsupials in Tasmania dating from about 23 million years ago, the beginning of the Miocene epoch, may indicate their contemporaneous presence in the forests and grasslands of the Lake Eyre basin.[13]

By the mid-Miocene, lakes and distributary streams dominated the area. The alternation of seasonal wet and dry intervals facilitated savannah grassland and shrubland expansion, home to the tapir-like marsupial herbivores known as palorchestids ('graceful leaper', named for the Greek hero Achilles' companion, Palorchestes, by Richard Owen). Forests with shrubby understorey grew along the permanent streams and lakes, in which abounded 'rayfinned fish, lungfish, chelid turtles, crocodiles and platypus' and species of estuarine dolphins, above which teemed a multitude of bird life – 'flamingos, ducks, pelicans, cormorants, gulls'. Fossil evidence also suggests a flourishing terrestrial mammalian fauna in the forests, including arboreal species such as gliding and ring-tail possums, burramyids, phalangerids and koalas, while the understorey provided niches for 'the small insectivorous dasyurids, bandicoots, wombats, rat kangaroos and the carnivorous thylacoleonids'.[14]

Panel 6

A marked cooling at the end of the Miocene preceded warming in the early Pliocene epoch, five million years ago, and cooling in the late Pliocene and early Pleistocene. Australia moved closer to Asia, facilitating biotic exchange between the two landmasses.

Four million years ago a warm shallow sea covered low-lying land in south-eastern South Australia. Sediments deposited here hardened into a fossiliferous sandstone, which is now exposed as a bluff along the north-western part of the Hallett Cove Conservation Park. Contemporary sediments in the Lake Eyre basin indicate increased fluvial activity, but as its northward drift continued, Australia dried. Subsequent lacustrine and alluvial deposits show a tendency to increasing salinity and suggest a seasonally arid regime. Saline shallow water sediments and gypsum crystallisation extensively overprint fossil evidence of the earlier freshwater lake ecologies (the presence of gypsum in a sediment is diagnostic of a hypersaline lake environment and generally indicates arid-phase geomorphological processes).[15]

This evidence is reinforced and mirrored by the changing terrestrial fauna and the disappearance of flamingos and freshwater dolphins. Erimean fossils of the early Pliocene epoch correspond more closely to the warmer, damper middle Miocene assemblage than to the later Pliocene. The latter fossil fauna is dominated by large browsing diprotodontids and the first evidence of large grazing and browsing kangaroos, which suggests the expansion of the open savannah grasslands and shrublands to something more closely resembling the landscapes of the region in the late Holocene, before the introduction of hoofed ungulates approximately 150 years ago.[16]

Elsewhere in 'South Australia', the present Mount Lofty Ranges were uplifted about two million years ago, initiating a new period of erosion in the 'Adelaide' region and around Hallett Cove. Across much of the rest of the continent, grasslands and eucalypt woodland gradually supplanted the previously abundant rainforests. Grazing mammals diversified, whose ancestors probably appeared in the early Palaeogene period, around 50 million years earlier (see Figure 21).

Panel 7

The pattern of warming and cooling, wetting and drying, with a tendency to increasing aridity, repeated throughout the Pleistocene (about 2.5 million to 11 500 years ago). Lake Eyre basin sediments reflect sea level and temperature fluctuations as the Pleistocene glaciations of Europe and North America swelled and declined according to oscillations of the polar ice caps. Full lakes, high stream discharge and marine incursions over the continental shelf appear early in the glacial cycle. Glacial maxima correlate with sea level retreat, dune building, dry lakes and the deposition of evaporite minerals like gypsum.

The last phase culminated around 20–25 000 years ago. Reduced precipitation and spring run-off from the east coast highlands made a dust bowl of the interior. As before, plains of sand blew across the inland to shape its characteristic dune systems. Sea level changes sculpted the coastlines as the continental shelf became coastal plain then shallow sea, and plain again. Giant birds and marsupials browsed the forests, grazed the plains and died in the mud of

Figure 21: Diprotodon fossil in Lake Callabonna, South Australian Museum expedition, 1893. (*In E.C. Stirling, 1900. The physical features of Lake Callabonna,* Memoirs of the Royal Society of South Australia, *1(2): 1–15. National Library of Australia Nq 560.9942 ROY (2053869))*

drying lakes (Figure 21). The fluvial and lacustrine faunas of the Pleistocene, which during humid cycles included 'fish, turtles, crocodiles, the aquatic bird life not dissimilar to that found today (e.g. pelicans, ducks, cormorants, herons, ibis, spoonbills)', show little change.[17]

In contrast, much of the large terrestrial fauna of the period appears to have become extinct well before the last glacial maximum. This includes the large goose-like birds such as *Genyornis*, giant browsing herbivorous diprotodontid and macropodid (kangaroo) marsupials, and the suite of carnivores, reptilian and marsupial, that probably preyed upon them like the giant lizard Megalania (*Varanus priscus*), the land crocodile, *Quinkana*, and the marsupial 'lion', *Thylacoleo carnifex*. From around 16 000 years ago to the present, the continent warmed by several degrees. Forest replaced the steppe grassland and open savannah woodland of south-eastern Australia. For at least the last 45 000 years, Aboriginal people have lived in and modified their customs and these landscapes according to the demands of climate and country.

Modern Lake Eyre is the fourth largest terminal lake in the world. Its shoreline is 1400 kilometres long, and at 15 metres below sea level it is the lowest point on the Australian continent. It is the topographic sump of the Lake Eyre basin internal drainage system, which, at more than one million square kilometres, covers almost one-sixth of Australia. The mean summer maximum in this area of South Australia exceeds 40°C and individual maxima as high as 61°C have been recorded. The region averages only 110–125 millimetres of rainfall every year and annual evaporation exceeds two-and-a-half metres.[18]

The arid landscapes of the basin apparently exceed in dryness even those of the late Pleistocene, when the fossiliferous deposits of Lake Callabonna accumulated. Sparsely vegetated plains and dunefields contrast with saltpans and ephemeral lakes and rivers, along which endures remnant woodland. In years of high rainfall in the catchment areas of the Diamantina, the Georgina and Cooper Creek (respectively in the Northern Territory and western Queensland) as in 1999–2000, 2007 and 2009, the refreshed watercourses flood Lake Eyre and an abundance of fish and waterbirds returns much as in the Pliocene, although the crocodiles, flamingos and dolphins are absent. In contrast, terrestrial faunas have undergone a revolution and introduced placental mammals dominate the country, with remnant populations of rabbit bandicoots, native rodents and red kangaroos. The new megafauna are hoofed ungulates like horses, cattle, sheep, goats and camels, and the largest carnivores are dogs, cats and foxes, which prey mostly on rabbits.[19]

These and similar narratives of degeneration are common in accounts of the interior of Australia. The first Australians had particular local versions which explained the presence and shape of particular landforms, fossils or processes. Some of these accounts, such as those of the *kadimakara*, giant creatures who came down from the sky to feast on central Australian verdure and became separated from their celestial home and died as the country dried, were recorded by Lutheran missionaries and ethnologists during the nineteenth century. They were common explanations for bones in the landscape and informed the speculations of nineteenth and early twentieth-century geologists. Other nineteenth-century versions, such as those recounted by Ralph Tate or by Edmund Charles Stirling, similarly linked the disappearance of the marsupial megafauna to a continent-wide drying, from a tropical paradise in 'the Diprotodon age' to the modern arid and semi-arid landscapes of South Australia. Tate assembled uplifted terraces in the Adelaide region, the glacial features of Hallett Cove and the Fleurieu Peninsula and the evidence for drying and extinction at Callabonna into a synthesising geotectonic hypothesis that accounted for the landscape features and faunal change on a continental scale. Anecdotal evidence often pointed to post-settlement drying and salination. At the turn of the nineteenth into the twentieth century, the Scottish-born Professor of Geology and Mineralogy at the University of Melbourne, John Walter Gregory, invoked both Aboriginal accounts and contemporary secular science. He famously wrote of a formerly lush 'dead heart' of Australia, whose potential might be reawakened.

Modern scientific versions involve climatic degeneration from a land of water to a land of dust. They parallel other secular and non-secular stories of ecological decline from a state of pristine, human-less 'nature'. Human agency in the landscape is often represented as negative and destructive, from the debates about causes of the extinction of Pleistocene megafauna to discussion of the limits of agriculture and pastoralism in modern Australia. Many of the same issues and lines of evidence preoccupied naturalists in the nineteenth century seeking to explain, disprove or validate arguments for climate change, extinction and human occupation in prehistory.

Chapter 5

'A WORLD PREVIOUS TO OURS', 1795–1892

Vertebrate palaeontologists have a keen sense of their own disciplinary history. They represent it both teleologically, as a progression from an ill-defined branch of the catholic natural philosophy of the early nineteenth century to the specialist science of the twenty-first, and adversarially, as having overcome the strictures of Church dogma, periodically cooperated with other evolutionary scientists to attempt to quell modern outbreaks of Biblical literalism, and struggled for relevance against the fluctuations of political, economic, religious, pedagogical and intellectual fashion. Several places are celebrated in the annals of Australian vertebrate palaeontology for their position in the chain of patronage, rivalry and negotiation that leads to the manufacture of palaeontological knowledge. Stories about these places provide some of the foundation myths of the discipline in Australia, notably the identification and naming of the *Diprotodon optatum* holotype by the great British comparative anatomist Richard Owen in 1838 (see Figure 22).

In the nineteenth century, the fossil remains of the giant extinct marsupials encroached on discussions about the origin of species, mechanisms for extinction and the antiquity of Aboriginal people. As part of the late twentieth-century and early twenty-first-century rehearsals of some of these debates, they have come to epitomise the destructive capacity of human beings in the often emotive, contentious question of the extinction of the Australian Pleistocene megafauna.

Nineteenth-century distribution studies and the age of the world

At the beginning of the nineteenth century, western European scientific understanding of geological time was an amalgam of two separate and apparently largely irreconcilable traditions: a Judaeo-Christian heritage which postulated a recent, finite age for Earth, calculated from Biblical chronologies in the Middle Ages to approximately 4000 BC, and Aristotelian-derived Eternalism. The Eternalism of Aristotle postulated that Earth/Nature is a 'primary entity', self-generating, self-sustaining and existing from all time. It found little favour in the Christian West before the eighteenth century, but leaves echoes in James Lovelock's late twentieth-century Gaia hypothesis. In contrast, Biblical scholars believed Earth and its inhabitants had been created, almost simultaneously, by Divine fiat out of nothing (Special Creation), and that geological and landscape features arose rapidly and catastrophically over a short period. During the seventeenth and eighteenth centuries, the rise of geological study undermined the scholarly tradition of Biblical literalism. The French naturalist Georges-Louis Leclerc, Comte de Buffon, deduced from experiments with heated metal spheres that he needed up to three million years to account for the cooling of the world from a molten ball. James Hutton argued in 1795 that the age of the world was unfathomable. It showed 'no vestige of a beginning, no prospect of an end'.[1]

The revision or redeployment of broadly accepted rules governing the genesis and behaviour of the natural world was partly a result of the development and increasing specialisation

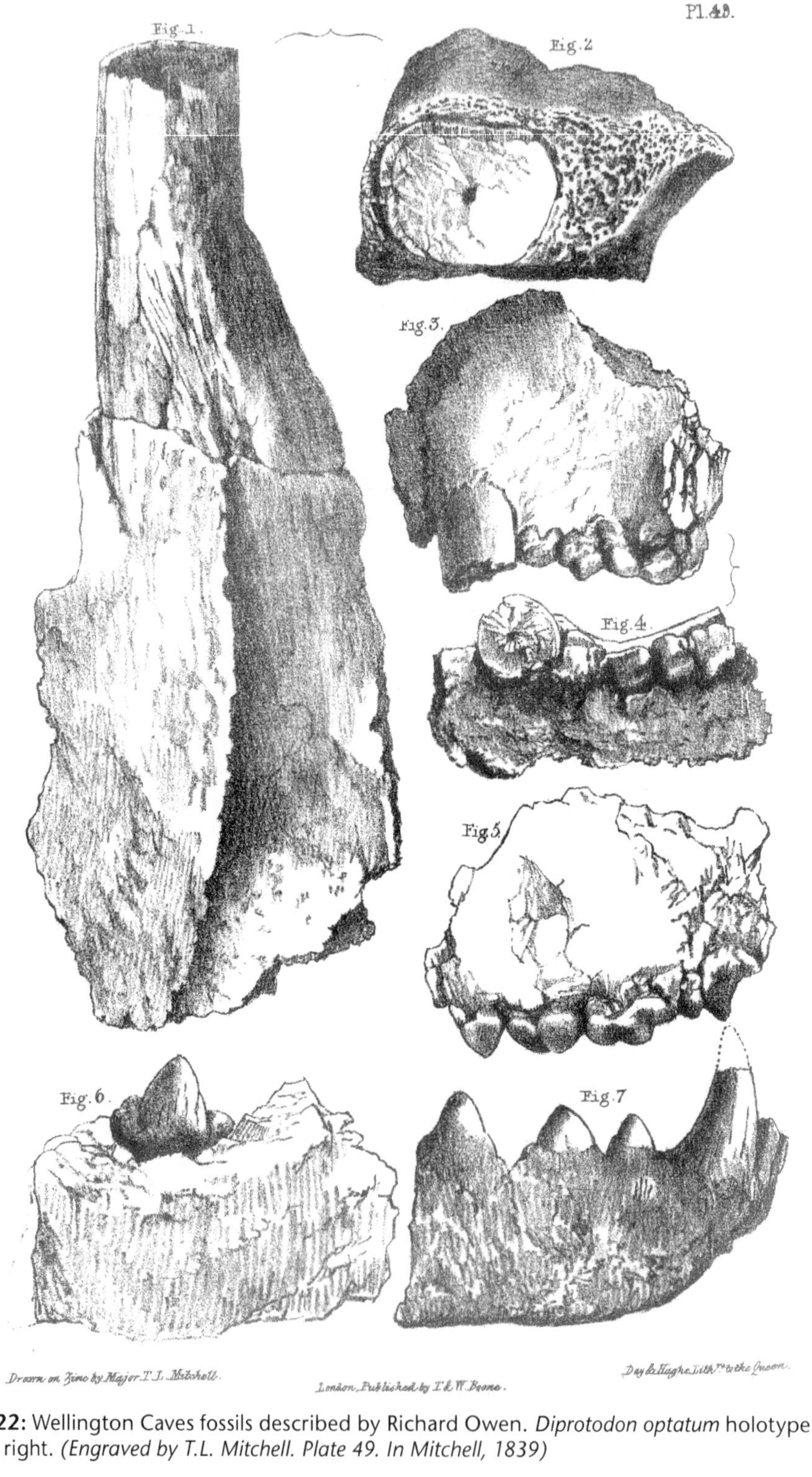

Figure 22: Wellington Caves fossils described by Richard Owen. *Diprotodon optatum* holotype top left and right. *(Engraved by T.L. Mitchell. Plate 49. In Mitchell, 1839)*

of geology from the department of natural history known as physical geography and also of discoveries in the New World. But it by no means involved a complete paradigm shift. Early in the nineteenth century, many high-profile geologists accepted that Earth must be millions of years old. Debate still raged over mechanisms for change, notably between those who approached the shape of the planet with a steady-state outlook, epitomised by the uniformitarian paradigm delineated in 1830 by Charles Lyell in *Principles of Geology*, and those like Georges Cuvier, the cleric and geologist William Buckland, and Louis Agassiz who thought that the current shape of the planet's surface was wrought by cyclical catastrophes – sudden cataclysmic events on a global or regional scale, the latest of which they often conflated with the Mosaic account of the Deluge.[2]

Debate broadly concerned distribution studies, the study of the distribution of living and fossil flora and fauna across the planet now termed biogeography. Unexpected fossil affinities were a sticking point. For example, the presence of the bones of fossil reindeer and fossil elephants (one species known to be cold-loving, the other tropical) in the same strata, or the presence of so-called tropical fossil assemblages in temperate regions puzzled naturalists. They called upon catastrophic flood, rapid climate change, migration, discontinuities in the fossil record, or the inscrutability of God to explain the apparent irregularities. They also keenly contested the association of fossilised human remains and artefacts with these fossils.[3]

These apparently anomalous relationships and other phenomena such as extinction – the disappearance of species that are represented in the fossil record but not found alive anywhere, such as the North American mastodons or the giant South American ground sloths, Megatherium and Megalonyx; and such cryptic presences in the landscape as the thick layer of sediment known as diluvium or drift which covers much of northern and western Europe – were often explained in terms of the latest of these cyclical catastrophes (a cataclysmic deluge or a sudden ice age for instance) or by invoking various mechanisms for gradual climate change across the planet (again including marine transgression or glaciation).[4]

As early as 1796, Georges Cuvier had argued convincingly for the reality of extinction in the fossil record. In a paper to the *Institut national*, he claimed that fossil evidence suggested powerfully an earlier, prehuman world or creation that had been destroyed by some kind of 'catastrophe'. He questioned how else bones resembling animals which are seen alive 'only in the tropical zone of the Old World' could be found in great numbers in the north of both Eurasia and North America when they 'cannot have belonged to any of the species that live today in those climates', also reasoning that animals as obtrusive as mammoths and mastodons could not possibly exist alive on Earth and yet remain unknown to science.[5]

Although many (but not all) naturalists were prepared to accept the difficulty of hiding a breeding colony of mastodons on an increasingly mapped and populated Earth, the situation with respect to flora, invertebrates and marine fossils was more problematic. Wandering naturalists around the world collected biological material which was formerly only identified in fossil form. For example the collections of the Baudin expedition in Australian waters (1800–04) included samples of the clam *Trigonia* dredged alive from Bass Strait but previously recognised only from Mesozoic marine sediments in Europe. With such discoveries and the presence of living marsupials allied in characteristics with the mammal fossils of the Oolites and Lias sediments of the Paris Basin (equivalent to today's Triassic and Jurassic periods) came the longstanding notion of Australia as a land of living fossils. As late as 1863 Hugh Falconer observed that: 'Australia has a living fauna, so low and backward in the scale of organization, that it has struck those who have reflected on it, in the light of being an arrested fragment of an older world, in which progress was suspended, whilst in the other continents it was being steadily sustained, by the appearance of higher and higher forms.'

Seven decades after Baudin's expedition, W.T. Bednall wrote in 1877 that the presence in Australia of a living *Trigonia* added 'another instance in which this great southern island–continent is the habitat of forms of life which in other parts of the globe' are only found as fossils. Cuvier's great rival and colleague, the transformist Lamarck, had earlier championed this idea. Lamarck denied the reality of extinction in the fossil record. He claimed that apparent extinction was in fact an artefact of the adaptation of species to changing conditions. The only true extinctions he ascribed to human activity during historical times. The importance of the Australian record as applied to the 'problem' of the disappearance of species and higher taxa was thus entrenched early in the history of comparative anatomy, evolutionary biology, and palaeontology.[6]

By Cuvier's death in 1832, the dramatic success of four decades of research on fossils had reworked his early demonstration of a single organic revolution into a powerfully heuristic palaeontological synthesis. Charles Lyell, the stratigrapher William Smith and other geologists established that the history of the planet, unimaginably long by the standards of human history, was documented by a thick sequence of slowly deposited layers of sediment. Successive formations demonstrated distinct assemblages of marine and terrestrial fossil species, enabling their widespread identification and correlation. This facilitated the development of a global geological time scale which relied on the premise that the history of life and the deposition of sediments had been, broadly speaking, the same all over the world. Thus, as with glaciation, the occurrence of such phenomena as the fossil remains of megafauna in the Southern Hemisphere was a point of great interest and debate, although unlike the glacial debate the question was not so much *whether?* as *what kind?*[7]

On 18 May 1831 Cuvier's colleague, the Irish naturalist Joseph Pentland, concluded a letter to Roderick Murchison at the Geological Society in London with the request that Murchison let him know 'what fossils you have received from New Holland. I see by the Literary Gazette that a paper on the subject had been read to the G.S' (Geological Society). The paper is probably Thomas Mitchell's 'Account of the limestone caves at Wellington Valley in New South Wales and of the situation near one of them, where fossil bones have been found', which appeared in the *Proceedings of the Geological Society of London* in April 1831. A paper by the cleric J.D. Lang appeared the same year in the *Edinburgh New Philosophical Journal* announcing the discovery of these same Wellington Caves fossils. The gradual discovery of the nature of Australia's extinct fauna played a part in the elaboration of that precocious and eclectic branch of natural history, distribution studies. It also enabled the correlation of Australian strata with those of the rest of the world.

The antediluvian Antipodes

The European intellectual colonisation of and by the Antipodes in this emergent intellectual landscape was well under way by the time Charles Sturt and Thomas Mitchell in the 1820s and 1830s traced Australia's landform history in the south-east back to a widespread or universal deluge. Both explorers invoked a colossal flood to account for particular features of the geomorphology and the distribution of fossil remains in the Murray–Darling basin. The landscape provided elucidation of God's word as drawn from Scripture. In its flatness, its 'island-like' chains of hills, its fossiliferous caves and cliffs, its lacustrine and fluvial geomorphology and its swamps and dunefields, the 'impoverished' material culture of its 'primitive' people and its 'anachronistic' flora and fauna, page after page revealed a quality of newness that engaged with guarded contemporary debate regarding the longevity of the planet and the literal interpretation of the Bible. According to both explorers, the landscape and its products as a system of signs could be inserted into their common literary, scientific and religious tradition to unlock the geological history of the country.[8]

Mitchell took an actualist approach. When John Oxley, his predecessor as surveyor-general of New South Wales, leading the 'Second expedition of discovery', arrived at and named the Wellington Valley in June 1818, he did not imagine he had discovered such fertile ground for the annalists of antediluvian worlds as well as for settlers, sheep and crops. Ten years later George Ranken discovered the first evidence that 'at some former period of its history' Australia was inhabited 'by various races of animals, that are either extinct or no longer existing in this part of the world'. Lang provided a detailed account of Ranken's discovery in the 'remarkable cave', announced first in the *Sydney Gazette* of 25 May 1830:

> On lowering himself ... Mr R observed, to his very great surprise, a piece of bone lying on the floor of the cavern ... on a more minute examination, he discovered a vast number of other bones... The rope by which he had lowered himself into the cavern had been fixed to what appeared a projecting point of the solid rock, but on its breaking off in consequence of the weight attached to it, it was ascertained to be a large fossil bone – the thigh bone, I conceive, of some quadruped much larger than the ox or buffalo, and probably of the Irish elk, the rhinoceros, or elephant.

Richard Owen later identified the 'thigh bone ... of some quadruped' as the tibiatarsus of an enormous bird, rivalling the moas of New Zealand in size. He eventually named this bird *Dromornis australis.* Palaeo-ornithologists now regard it as a precursor to *Genyornis newtoni,* a Pleistocene giant first identified from Lake Callabonna specimens in the 1890s.[9]

Mitchell visited the Wellington Valley in late June or early July 1830 accompanied by Ranken, to survey the caves and collect fossils. He recorded in 1839 that the bones suggested powerfully 'a deposition from water' having earlier laid out his evidence for these and other 'marks of various changes in the relative height of the sea' on the east coast 'and in the interior'. In a characteristic disclaimer, he added that he could not 'pretend to account for the phenomena presented by the caverns' and then proceeded to try. Based on the geological and geomorphic evidence which he laid out in his exploration journals and field notes, Mitchell hypothesised that the plains of inland New South Wales 'were formerly arms of the sea: and that inundations of greater height have twice penetrated into, or filled with water, the subterraneous cavities' of the Wellington Valley and other caves in the colony.

It is not clear whether he thought that the inundations had a direct causal relationship with the fossil animals found there. But his expeditions took place well before Agassiz formulated his glacial theory accounting for the superficial deposits of Western Europe, and Mitchell was a disciple of William Buckland's diluvial theory. In 1830, before venturing to the Wellington Caves fossil site with Ranken, he expressed his intention of finding 'Antediluvian Remains, like those found by Mr Buckland'.[10] As late as 1850, heading to the Wellington Valley on a mineralogical survey, he wrote to Richard Owen on 13 April that '"*reliquiae diluvianae*" shall not be overlooked'. As he wrote again to Owen on 5 January 1852, he was forced to neglect the study of *reliquiae diluvianae* after all, as the urgency of his 'mission – and want of assistant hands', and perhaps a gold-fixated colonial government, refused to extend his researches from the auriferous rocks of the Wellington Valley back to the limestone caves.

In line with Buckland's and Cuvier's theories about successive cataclysmic events separating different faunas representing different lost worlds, the Wellington Caves evidence suggested to Mitchell that flooding separated two distinct and unrelated fossil assemblages. As he wrote in *Three Expeditions* in 1838, the 'operations', or cataclysmic events, appeared to have occurred subsequent to the local extinction 'of the species whose remains are found in the breccia; and previously to the existence, in at least the same districts, of the present species'. Mitchell was even less ambiguous about this intellectual legacy in his field notebook. He wrote on 14 October 1830 that the caves show 'the same proofs of a deluge, which are found in the caves of Europe', describing 'a deep deposit of such particles of earth' as 'could have been easily suspended in muddy water and carried downwards from the surface through fissures'. The first version reads 'proofs of *the* deluge' but 'the' has been struck out and replaced with the indefinite article.

Citing Mitchell's data, his own observations and contemporary logic, Charles Sturt proved to himself in 1833 'the comparatively recent appearance above the ocean of the level country' he had charted in south-eastern Australia. By his reading, Australia was the last continent to emerge from the retreat of floodwaters 'which swept the vast accumulation of shells, forming the great fossil bank of the Murray from the northern extremity of the continent, to deposit them where they are'. He concluded that 'the depressed interior' must simultaneously have been covered by this inland sea which 'washed the western base of the dividing ranges'. Most importantly, the discoveries of stratified fossil deposits at the Wellington Caves confirmed the newness of the country and the existence of at least one past world.

The close of the second volume of Sturt's 1833 journal was therefore an attempt to synthesise the geographic evidence constantly rehearsed in his published narrative for the post-diluvial quality of the eastern Australian landscape. He observed fossil-rich cliffs along the Murray in 1829 and masses of oyster shells in the same vicinity on the tops of hills, as well as an interbedded sandstone and clay formation upstream that seemed 'rather the work of art than of nature'. Changes were no accident but the consequence of direct agency. Geological research daily revealed more signs of the 'awful catastrophy which so nearly extinguished the human race'. These marks and changes were 'as gleams of sunshine falling upon the pages of that sublime and splendid volume' which hitherto stood 'single and unsupported before mankind'. The corroborative testimonies of God's wrath were not known to 'those who wrote that sacred book' and so 'the discovery of the remains of a past world' must strike Sturt and his readers with the 'truth of that awful event', the Deluge of the Mosaic account.

The footprint of the Mosaic account was lighter in his journals of the central Australian expedition (1844–46), in his field journal (edited by Richard C. Davis in 2002 for the Hakluyt Society as *The Central Australian Expedition*), in the relatively candid and often despairing weekly journal he wrote to his long-suffering wife Charlotte while in the field (published in 1984 as *Journal of the Central Australian Expedition*) and in the published version of 1849 (*Narrative of an Expedition into Central Australia*). Although his sense of predestination, his appeals to Providence and his conviction of the watery genesis of the central Australian dunefields were undiminished, his accounts at the beginning and end of the 1849 publication of the former shape of Australia, wherein he speculated that the continent was once an archipelago, are attempts at a geological, rather than a Biblical, synthesis.

The middle years of the nineteenth century had continued the slow revolution in the investigation by naturalists of changes to the planet's surface. Perhaps in the intervening years Sturt read Lyell's *Principles of Geology* or Charles Darwin's acclaimed third volume of the official narrative of the *Beagle* voyage, reprinted in 1839 as *Journal of Researches*, or Darwin's contributions to the nascent science of geodynamics – the two monographs on island formation which sprang from his observations on the *Beagle* voyage, *Coral Reefs* (1842) and *Volcanic Islands* (1844). In any case, the work of Darwin or something like it informed Sturt's later accounts of the lost worlds of the interior of Australia. He was certainly no clearer about the mechanism for the floods that 'must have swept' across central and southern Australia to produce the 'unparalleled desert' that nearly killed him, but he no longer explicitly identified it in print, even to his wife, with the 'awful event' which 'so nearly extinguished the human race'. In keeping with Sturt's and Mitchell's theories, George Ranken's discovery of the fossil remains of the Wellington Valley provided an opportunity for J.D. Lang to bring Australia geologically into line with the Old World with 'another convincing proof of the reality and the universality of the deluge'. Thanks in part to the Wellington Caves fossils, Lang was able to reinscribe supposed signs of youth in the landscape as signs of age comparable with Europe. By analogy with William Buckland's identification of an antediluvian hyena den at Kirkdale in Yorkshire, Lang declared that 'this vast island is not of recent or post-diluvian formation, as is generally asserted, without the least shadow of evidence'.

The discovery inspired Lang 'with a powerful motive of gratitude to Divine Providence for that long-forgotten visitation' which, by destroying 'such beasts of prey as the antediluvian inhabitants of the cave at Wellington Valley' – elephant, hippopotamus, Irish elk, hyena, tiger, rhinoceros – created 'so eligible a place for the residence of man'. It also provoked a lament that 'the interesting youth of Australia' should 'hitherto have been debarred, in consequence of their want of instruction in the various branches of Natural History and Natural Philosophy, from prosecuting the numerous and interesting paths of discovery which this vast island presents to every man of science and research'. He hoped that by promoting abroad the natural history of Australia's deep past, he could also further the cause of Antipodean universities.

These early British annalists of Australia's former worlds discovered unpeopled geographies in the dirt and rocks. It was not long before related finds of scratched bones, tools and mysterious teeth began to impact on debates about antiquity of human beings and the origins of the Australians and the Tasmanians, which continue or are revisited today. Australia was a fabulous bestiary of anachronistic natural science specimens to hunt, shoot, stuff, dissect and pickle, and when Lang took a package of fossils from the Wellington Caves to Professor Robert Jameson in Edinburgh in August 1830 it became a scientific necropolis as well, yet another site of transcontinental renown for the investigation and imagination of lost ecologies, like Big Bone Lick in Ohio, Kirkdale in Yorkshire or the Cavern of Lunel in France.[11]

Several British savants, notably Charles Darwin, Joseph Hooker, Thomas Huxley, and Richard Owen, spent a significant part of their careers in the consideration and classification of the palaeobotanical, geological and palaeontological remains of the Antipodes and fostered wide networks of Antipodean collectors and correspondents. In 1833 Charles Lyell referred to the Wellington Caves remains as supporting non-Cuvierian principles (notwithstanding the later publications of Charles Sturt and Thomas Mitchell) and the significance of the marsupial nature of the Australian fossil mammal record has been implicated in Darwin's development of the law of succession of types or descent with modification.[12]

'Herbivourous mammals of such magnitude'

'A Gigantick species of Kangaroo'

Mitchell sent a second batch of bones to England in September 1831, including a mysterious tusk-like tooth. These, along with 'a copy of the description and some drawings' were intended for Cuvier, 'from a desire to afford some information respecting the fossil remains we have found in New South Wales' and 'in hopes to profit in common with the rest of the world' from the Baron's anatomical wizardry.[13] Unfortunately, the French naturalist died in 1832 before the bones reached him, but he had previously seen some of Mitchell's drawings.

Both collections found their way to Edinburgh, where they were examined by Jameson, and to London where William Clift, the conservator at the Hunterian Museum of the Royal College of Surgeons, and Pentland, Cuvier's associate, examined them. Pentland also appears to have seen the bones sent by Lang in 1831, which suggests Cuvier probably saw them before he died. Pentland wrote to Roderick Murchison from Paris on 1 May 1831 that he had 'received some new fossils, from the New Holland Caves. They belong to a Gigantick species of Kangaroo, 1/3 larger than the largest specimens of the largest Macropus of which we possess skeletons – and in all probability to another extinct species – this is the 8th species found in these caverns.' Between 1831 and 1833 the Irish naturalist published at least three papers on the Australian fossil remains, translated between French, English, and German.

Mitchell read widely. In 1830 he was able to write of the Wellington Caves fossils that he had 'been much enlightened by the able works of Professor Buckland and M. Cuvier'. He had read enough to remain puzzled: 'still the circumstances of these bones seem to me quite inexplicable ... I shall feel highly gratified if these specimens of Fossil bones from this side of the Globe, engage the attention of any such distinguished geologists.' Pentland's prolific notices suggest that the specimens did indeed generate interest.

Apparently anticipating Darwin's law of the succession of types, Mitchell mused that the existence of that 'singular animal', the kangaroo, at such an 'early age' in Australia 'may lead us to suppose that the other remains belonged to species never known in other countries – and it is highly interesting to ascertain therefore whether this is the case or to what species of animals such bones belong'.[14] However, in a letter to Richard Owen in January 1843, assuming the Diprotodon to be a gargantuan kangaroo, he clarified his position. It was 'consolatory' to find that the southern continent 'did once support herbivorous animals of such magnitude', in keeping with Eurasia and the New World. Evidence like Mitchell's, of 'an animal so well provided for a country of burning woods and fallen timber – by its young-protecting pouch and saltatory powers' having 'always belonged to Australia', might have suggested to Darwin the idea of descent with modification.[15] But to Mitchell it suggested the action of providence and the immutability of species which are, in the words of the Melbourne-based German botanist Ferdinand von Mueller, 'clearly defined in nature, all perfect in their organisation, all destined to fulfil by unalterable laws those designs for which the power of our creating God called them into existence'.[16] Momentarily preoccupied by 'the curious gradation of species – and the diminutive character of existing classes' which 'seem to indicate the energies of animal nature here to be on the wane', Mitchell chose to dwell instead on the 'wise provision of providence for the introduction of those other large animals by man's agency – which have been found better suited to his wants'.[17]

On a visit to England in 1838 to oversee the publication of *Three Expeditions*, Mitchell took further Wellington Caves material and a partial jaw found in the channels of the Darling Downs to Owen, whence began their fruitful correspondence. Owen, recently elected the Hunterian Professor, and now William Clift's heir apparent at the Hunterian Museum, was set to make the study of Australian vertebrate fossils a significant component of the rest of his career.

The first unprepossessing but fateful metropolitan recognition of the nature of *Diprotodon optatum* – named for its two front teeth – appeared in a letter from Owen which Mitchell reproduced at the end of the second volume of *Three Expeditions*. The letter is dated 8 May 1838 and differs little in substance from the published version.[18] Owen's classification is based on a piece of characteristically inspired deduction by the anatomist who had only 'the anterior extremity of the right ramus, lower jaw, with a single large procumbent incisor' from which to draw his conclusion (see Figure 22, page 88). Jameson and Clift initially identified the animal as a hippopotamus or dugong; Buckland and J.D. Lang thought it likely to be a rhinoceros or hippopotamus; Cuvier and Pentland suggested it was a young elephant. Other later suggestions included sloth, tapir, kangaroo, koala and extinct pachyderm. The Port Phillip-based naturalist, E.C. Hobson, identified the 'incisor of an enormous rodent' among the remains found at the Lancefield Swamp near Mt Macedon in 1838, which Owen later identified as a Diprotodon tooth. In 1844 Ludwig Leichhardt took issue with Owen's classification of Darling Downs dental material (which might or might not have been Diprotodon) as belonging to a 'gigantic pachyderm' allied to the *Deinotherium*, a genus of extinct proboscidean placental mammals found in sediments from lower Miocene to upper Pleistocene, distantly related to elephants. As he wrote to Owen on 10 July 1844, Leichhardt assumed it was Diprotodon material and insisted it was more closely allied to kangaroos than 'any other animal'.

Owen was the first naturalist to recognise the mystery animal's marsupial affiliations. He outlined his philosophy on comparative anatomy to Roderick Murchison in 1845:

> It is only by the knowledge ... of existing species that we can possibly arrive at that of the extinct ... [We] must not infer from one species of Elephant that another had the same identical kind of food. True; but it is so only by knowing what kind a living species *has*, that we can know the kind & degree of deviation which accompanies specific distinctions of dental structure in an extinct species.[19]

His classification was based on a comparison with the wear, position and structure of a wombat tooth and with the teeth of some of the placental mammals to which it had already been assigned. He wrote to Mitchell:

> This [Figure 22] is the specimen conjectured to have belonged to the Dugong, but the incisor resembles the corresponding tooth of the wombat in its enamelled structure, and position ... But it differs in the quadrilateral figure of its transverse section, in which it corresponds with the inferior incisors of the hippopotamus.[20]

Still, even Owen could not define the fossil's marsupial associations any more precisely, speculating that it might have been a kind of kangaroo, a giant wombat, or even a relative of the koala.

'One incisor of an enormous rodent'

The Mt Macedon site, near the Victorian town of Lancefield, has the distinction, after the Wellington Caves and the Darling Downs, of being the third megafaunal site discovered by Europeans in Australia, the first in the Port Phillip district. It might also be the first site discovered containing a species of the birds which the Monash University palaeontologist Patricia Vickers-Rich has nicknamed *mihirung* birds. These are the *mihirung paringmal* or 'giant emus' of the Tjapwurong people of western Victoria. Vickers-Rich suggested that the 'birds as high as mountains' might correspond to the giant flightless birds of which *Genyornis newtoni* is the most recently extinct. She speculated that the stories of the mihirung were evidence not just of the anatomical skills of Aboriginal people who identified the fossils as belonging to giant birds, but that they recalled a time when Tjapwurong and mihirung coexisted in the volcanic plains of Victoria's Western District. Both the mihirung and active volcanism fill out the narratives. Volcanoes have been active in the Western Districts as recently as 3000 years ago.[21]

Hobson listed the bones of 'some large cursorial bird, probably an Emu' among the Lancefield Swamp discoveries, but they were not formally identified. Australia's first scientific journal, the short-lived *Tasmanian Journal of Natural Science, Agriculture, Statistics &c*, published Hobson's findings, often via extracts from letters to his friend, the botanist and editor of the journal Ronald Campbell Gunn (including an extract which Gunn cheerfully admitted that he fabricated in a letter to Hobson in 1847, for the purpose of introducing two species of wombat).[22] Hobson mentioned the emus and 'gigantic birds' in several of these letters. Despite his close acquaintance with Richard Owen (Hobson had studied under Owen and Robert Grant at the Royal College of Surgeons in the 1830s) and the warmth with which they addressed each other, the only specimens from Lancefield to reach England were a few Diprotodon (the 'incisor of an enormous Rodent'), wombat and *Macropus titan* remains.[23]

The bones of the 'large cursorial bird' disappeared after Hobson's early death from tuberculosis, thereby, at least in the view of some commentators, potentially depriving Lancefield of the type locality for *Genyornis newtoni*. Historian of science Wayne Orchiston and geologists Rob Glenie and R.N. Miller speculated that the emu Hobson found was in fact *Genyornis*, as the bones of the enormous bird were found at Lancefield during late twentieth-century excavations. Instead it was left to Stirling and Zietz in the 1890s to identify the *Genyornis* holotype at Lake Callabonna, which Stirling named for his friend and mentor, Cambridge professor Alfred Newton.[24] As the bones were lost before anyone other than Hobson examined them, it is possible that they were indeed emu bones as he suggested. It is probable that he did not send them to Owen because he knew or assumed that ordinary emu bones would be of little interest to the busy comparative anatomist.

During the 1830s to the 1850s, the Wellington Caves and Liverpool Plains of New South Wales, Lancefield Swamp and Queensland's Darling Downs produced a macabre menagerie of extinct megafauna remains. Although W.B. Clarke wrote in 1847 that 'such relics of the former occupants of New Holland may be expected in all places where the physical conditions of the country appear to be of similar kind', these four regions remained the primary sites for the excavation of Australian megafauna during the nineteenth century, until the discovery of the Lake Callabonna site in the 1890s. However, vertebrate palaeontology was far from flourishing in the colonies. The 'expense & trouble of getting these things from the far Interior' was considerable, and pastoralists had other priorities. They often regarded fossils as curios, if aware of them at all, and left them to disintegrate, unless suitable rewards were forthcoming. Inaccessibility, ignorance and isolation remains a problem for palaeontologists, but before camels and the railway opened up the inland the difficulty of obtaining specimens from remote locations was incomparably magnified.

The Diprotodon's toe

The bone king

Labelled an 'English Cuvier' in the 1840s, 'the most distinguished comparative anatomist of the day' in 1847 and 'the greatest Anatomist of the Age' in 1870, Richard Owen's empire of fossil behemoths now stretched across several continents, geological periods and three vertebrate orders. Despite his preoccupation with such exotica as the giant terrestrial, marine and flying 'lizards' found much closer to home in the Sussex Weald, at Stonesfield near Oxford and in the cliffs of Lyme Regis, the Australian mammal remains continued to exert their pull. On 16 July 1846, he laid before the Board of Curators of the Hunterian Museum a report on 'the fossil bones of Quadrupeds recently received from Australia'.

Owen had previously produced a *Catalogue of Fossil Mammalia* incorporating 'figures of some of the fragmentary fossils of gigantic extinct marsupial animals in Australia', published in 1845.[25] This led to 'an active search after similar remains, and [to] the transmission to the College, on the parts of Lieut. Col. Sir Thomas Mitchell, Dr Hobson, Dr Leichhardt and George Bennett Esq., of the specimens so collected'. He set about selling to the board his plan to 'publish descriptions, with figures of the principal specimens, as a supplement to the volume of the Catalogue of Fossils'. Their 'novelty and interest' was such as to encourage him to produce a special report by which he hoped to gratify the gentlemen whose liberality and interest had so enriched the collections of the College.

The material consisted of bones gathered largely from the Wellington Caves (New South Wales), the Darling Downs (Queensland) and 'the District of Melbourne, 80 miles SW of the town of Melbourne' (Lancefield Swamp). Mitchell's contributions came principally from the Wellington Caves and Darling Downs, Hobson's from the Port Phillip District/Lancefield ('District of Melbourne') and George Bennett's from the Darling Downs. There were gigantic species of kangaroo, 'a characteristic fragment of the lower jaws of a Wombat of, at least, four times the size' of modern wombats, and most strikingly, extensive remains of 'the still more gigantic extinct pachydermoid Marsupials indicated in the fossil catalogue under the names of Diprotodon and Nototherium'. By this time Owen had renamed *Diprotodon optatum* the less whimsical but geographically evocative *Diprotodon australis* and the collections contained examples of 'the entire dentition of the lower jaw, in both young & mature individuals', as well as a portion 'of the upper jaw w/ molar teeth & an incisor, nine inches in length & one & a 1/2 inch in breadth' from the Lancefield Swamp (Hobson's 'incisor of an enormous Rodent') and other parts 'strikingly illustrative of the colossal size'.

The Australian remains were of great interest in their own right, for the 'very remarkable anatomical characters' they preserved, but even more importantly, for their comparative value, or 'the additional illustration which they afford of the geographical distribution of extinct Mammalia'. Found in 'the breccia of bone-caves [and] in fresh-water deposits of, apparently, the post-pliocene, or newer tertiary period', the fossils were referable to the 'formations containing the remains of mammoths, Hyenas, &c, in Europe', and thus were invaluable for trans-continental correlation. Furthermore, the specimens themselves were mostly 'peculiar in Europe to the Collection of the College', which, together with their 'novelty and rarity', made their publication a priority.[26]

Owen eventually had his way and produced first an Australian supplement to the mammal catalogue in 1845, and in 1877 his then definitive *Researches on the Fossil Remains of the Extinct Mammals of Australia*. But even by 1877, with a near-complete skeleton available, Diprotodon feet remained elusive, so Owen was compelled to represent his marsupial ankle-deep in grass. Nor had he identified Diprotodon tail bones or epipubic bone, the former suggestive of its closest marsupial relations, the latter conclusive proof that it was marsupial (see Figure 23).

'how it is that no large digits should occur'

Comparative anatomists required an entire Diprotodon skeleton to establish its affiliations. The discovery of a near-complete skeleton of a Diprotodon in the Darling Downs in the mid-1840s was extremely important in this respect. Hitherto known only by a few scattered teeth and tusks and sections of lower jaw, this skull was 'not only unique but perfect; for the head is nearly complete, with the exceptions of some of the hinder parts'. As Clarke explained, it exhibited 'all that is necessary for comparison with the Wombat, Tapir, and Kangaroo, and those gentlemen who have examined it, have pronounced it in many regards nearer perhaps to a Kangaroo than to any other known creature'. From their respective dental formulae and 'other reasons' that Clarke regarded as unnecessary to enumerate, the Sydney intelligentsia ruled that the Diprotodon 'must have been in some respects very near a Kangaroo, and if characters may be relied on, certainly marsupial', but as there was also 'a great resemblance' between its teeth and koala teeth only 'the presence of a tail would settle that point at once'.[27]

In contrast to these mid-century speculations, in 1869 Gerard Krefft, the curator and secretary at the Australian Museum from 1861–1874, agreed with Owen in classing the gigantic extinct marsupials *Diprotodon*, *Zygomaturus* and *Nototherium* with the phalangers rather than

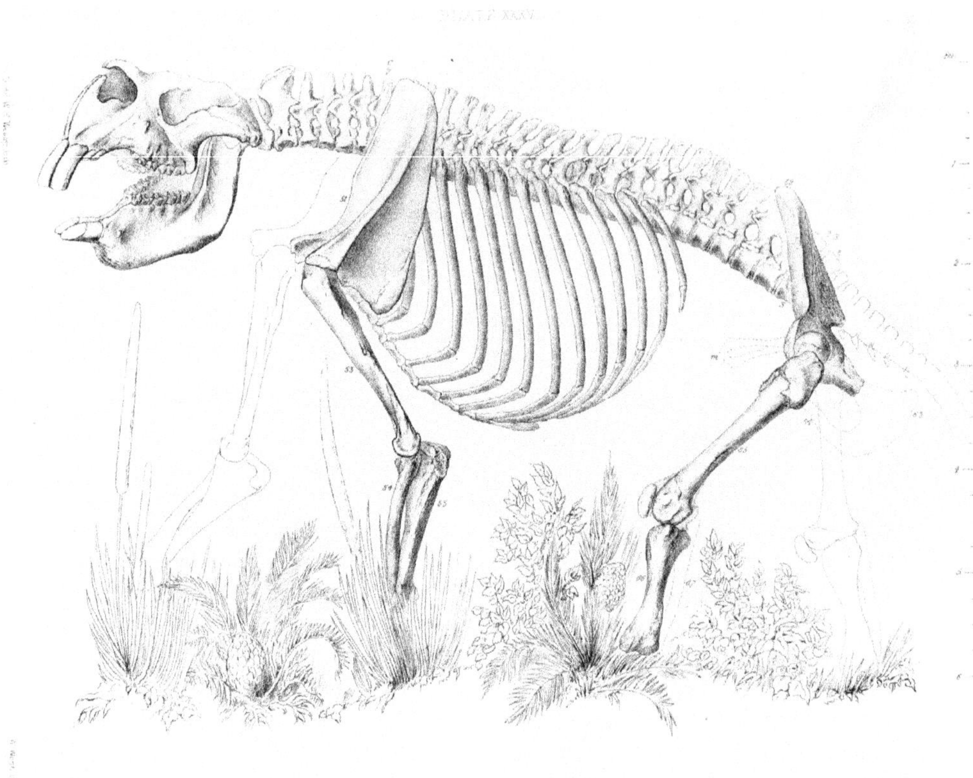

Figure 23: Diprotodon skeleton standing in grass. *(R Owen. Plate 35. In Richard Owen, 1877.* Researches on the fossil remains of the extinct mammals of Australia: With a notice of the extinct marsupials of England *(J. Erxleben: London). National Library of Australia NLq 569 OWE (382767))*

the kangaroos. As he wrote to John Robertson, the Colonial Secretary, on 10 May 1869, their incisors 'resemble the teeth of the phalanger more than those of the kangaroo' and their lower jaws 'are deficient of the wide excavation at the base of the coronoid process so conspicuous in the jaws of all kangaroos'. Still at question was the nature of Diprotodon feet. The missing bones were a crucial part of the biological jigsaw. Krefft asserted to Robertson that their structure would probably be found 'to have resembled that of the still living phalangers, such as the *Phascolarctos* [koala] and *Phalangista* [possums and gliders], or the wombats'. Krefft wrote in puzzlement to Owen's friend and collector, the physician George Bennett, on 23 February 1870 of the strangely digitally depleted Wellington Caves collections of the Australian Museum:

> I quite understand Professor Owen's surprise regarding the absence of the Carpal & tarsal [finger and toe] bones of the gigantic Marsupials. I often wondered what had become of them. They are missing, that is a fact ... I cannot understand how it is that no large digits should occur as these bones are generally very common & always well preserved even those of the smallest *Dasyures*. But so it is; they are still missing. You may imagine that I looked for them so that I am as much disappointed as our learned friend.

Later that year Krefft made an intuitive leap based on the structure of wombat feet and the absence of 'large phalanges [finger] & metatarsal [toe] bones' from the Australian Museum collection. Writing to Owen in June 1870 with his usual mixture of self-confidence and obsequiousness, he enclosed 'one of the flat phalanges, which I suppose to be that of the thumb of a gigantic Phalanger, probably a Zygomaturus or Diprotodon' and 'the last nail-less digit, which may probably belong to it'.

Owen had also noticed the lack of giant digits corresponding with the fossils of the diprotodont marsupials. Krefft mused that as Owen was aware that 'the nail-bones whereof I sent you the photographs are much curved & compressed & different from Kangaroo nails', he may also have observed that 'the fossil metatarsals are short & thick & that it is therefore not impossible, considering that both Diprotodons & Nototheriums (or *Zygomaturi*) stood rather short on their legs & progressed like Wombats or Phalangers, that these animals had hind feet like Phalangers, four nailed toes & a nail-less thumb'. He congratulated Owen for his early classification of Diprotodon 'with the Wombats', a 'more correct position than the persons who would make a sort of Kangaroo of it'. Owen based his assumption on the comparison of teeth and jaws, as did Krefft in 1869, but by 1870 Krefft believed that the answer lay in the mysterious tarsal and carpal bones in the collection: 'in this way only can we account for the absence of large metatarsals'.[28]

Krefft was a competent anatomist, but also appeared erratic and alcoholic. He fell out with the trustees of the museum and was controversially dismissed in 1874.[29] He also fell out with Owen in the early 1870s, engaging in a sustained attack on the British anatomist's identification of *Thylacoleo carnifex* as a lion-like marsupial carnivore. Krefft classed Thylacoleo with the phalangers and insisted that it was really no more than an overgrown possum, at best insectivorous, that might have used its blade-like incisor for stabbing melons and its 'formidable pre-molar' for 'bruising' its food.[30]

Perhaps Krefft's tarnished reputation and his soured relationship with Owen contributed to the neglect of his identification of Diprotodon feet in the Australian Museum collection. Whatever the cause, it was nearly two decades before he was vindicated by John Meldrum's parcel of bones from Lake Callabonna, delivered to the Black Swan Hotel, 11 years after Krefft died from respiratory failure.

Chapter 6

FINDING LAKE CALLABONNA, 1892–1901

Vertebrate fossil remains were an intellectual and material resource in nineteenth century Australia. The invention of Lake Callabonna as a significant South Australian landscape parallels an understanding of the past as a political and economic resource for colonial institutions.

It was in this environment, of overwhelming scientific and public curiosity about the material remains of South Australia's past, that John Meldrum gathered a bundle of Lake Mulligan bones and rode as hard as he could for Adelaide, and into notoriety. Simultaneously robbing Lake Mulligan of its hard-won individuality and Jackie Nolan of the sharp eye and enquiring mind that might have encouraged him to show the bones to an interested white man, the *Register* reported on 4 November that Meldrum, a 'well-sinker ... from Ragless's Calaburna Station, on the eastern side of Lake Frome' arrived in Adelaide the previous Tuesday evening, bringing with him 'some interesting fossiliferous specimens of the gigantic mammal Diprotoden [sic] Australis' which he claimed to have 'found on the shores of the lake, about four miles from the head station' where 'the surface sand had been blown away, revealing the bones referred to and others of larger size', too difficult to transport on horseback. The *Register* explained that the primary interest of these bones, which suffered 'in the manner of transport and exposure to the air' on the long trek to Adelaide and therefore had limited intrinsic value, lay in 'the proof of the existence of the giant, extinct marsupial of centuries ago in a locality in which its fossil remains have not hitherto been discovered', despite the criss-crossed tracks of 50 years of exploration, surveying and pastoralism in and around the locality. Once again, millennia of human husbandry were written away in a few words.

Having searched the lake all day, apparently Meldrum did not look far once he found the fossil site, simply grabbed as many fossils as he could easily carry. But he did report bones of three carcasses exposed on the surface near the specimens he collected, with 'joints of some of the femoral or thigh bones ... as large as an ordinary-sized basin'. Alert to the promise of fossil wealth, the *Observer* ran the article as well. Such was the excitement that the paper urged that 'Dr Stirling and Mr Zietz be laid on their track, as this interesting discovery is coequal with that recently made by our Government Geologist on his recent trip to the North'. On this trip in 1892, the Government Geologist in question, H.Y.L. Brown, found Pleistocene turtle and crocodilian fossil remains in the sediments of the Cooper and Diamantina associated with Indigenous *cadimurka* or *kadimakara* traditions. Indeed, the necropolis quickly eclipsed the Cooper Creek assemblages: on 13 June 1893, months after Meldrum's arrival, the *Register* reported that 'Mr Brown says that this is certainly the most remarkable discovery of fossil remains ever made in Australia, and perhaps the most singular ever made anywhere.'

Notwithstanding its singularity, the newspaper still reported the locality as 'near Lake Frome'. The Adelaide press compared the discovery of the Lake Mulligan remains to the Wellington Caves discoveries of 1829–31 for international scientific significance, and again the *Register* confused the location of the latter (well south of the Liverpool Plains), the timing of Mitchell's expedition (1830, 10 years later than the newspaper claimed), and wrote George Ranken out of the story, insisting that:

> Since Major (afterwards Sir Thomas) Mitchell, the explorer, found fossiliferous remains in the Wellington Caves, Liverpool Plains, New South Wales, some seventy years ago, to no other similar discovery has been attached more interest by the scientific world than to that of the giants now being exhumed by our Museum authorities on the shores of Mulligan Lake in this colony.[1]

The museum sent collectors to the site for reconnaissance. The board appointed Henry Hurst, formerly of the Queensland Geological Survey and the Queensland Museum, to lead the first expedition team. John Meldrum found employment there too. In a letter to the Board of Directors of the Museum, Public Library and Art Gallery written 14 months later (11 January 1894), Meldrum explained that when he first arrived in Adelaide in November 1892 he had showed the bones to a 'Reporter of the *Advertiser* who published a report in that paper'. Henry Hurst, one of 'several gentlemen' who inspected the specimens, took them to Stirling at the museum, which probably contributed to his eventual appointment as lead geologist of the museum's first expedition to the lake. A month later the *Register* and the *Observer* announced the appointment of Hurst and Meldrum, as reported in the minutes of the museum committee meeting of 2 December 1892. Mr Hurst was to be paid a fee of £20 and travelling expenses and Mr Meldrum was to be paid £3 per week and travelling expenses 'to bring away for the Museum what may be worth removal'. Any further reward was 'to be dependent on the value of the find'. Stirling, who chaired the meeting, reported that Hurst and Meldrum arrived at Leigh's Creek on 30 November and started for 'Lake Frome' the next day.

At the next board meeting two weeks later, on Friday 16 December, Stirling 'reported on the arrangements made for examination of the Lake Mulligan fossil remains and their securing by the Museum'. From Hurst's initial reconnaissance, it seemed the deposit was 'much larger than was supposed, and that its examination would require a larger party and a much longer time than had been originally contemplated'. The delighted committee agreed to consider the matter of a more extended search. They estimated the cost for a three-month stay at £250, which 'with strict economy' from the committee and free transit provided by the Railway Commissioners, they could make available. Hurst and the second excavation party departed Adelaide and arrived at Quorn on 13 January 1893. They left Farina on 20 January, proceeding 200 miles (320 kilometres) in eight days to Lake Mulligan. The party consisted of Henry Hurst, his brother George, John Meldrum (hired as excavator), 'two labourers and an Afghan' cameleer. The unnamed labourers were probably the elder Hurst brother (William, later described by Stirling as 'a sort of hanger on of the camp') and the other excavator, Templeton.[2]

Fred Ragless sets the story straight

It took some time for another story of the discovery of the remains to emerge.

In July 1893, pastoralist Fred Ragless wrote to 'disburden the public mind of the impression that Mr Meldrum was the discoverer'. While Meldrum might first have brought the bones to public notice, Ragless wrote rather sarcastically that 'unless the fact of his having borrowed a horse' from Callabonna Station and followed Fred's tracks to the spot 'eight days after I had been there' would 'entitle him to rank as the discoverer', the well-digger had assuredly not found the fossil site first. He explained why the busy Callabonna pastoralists took so long to refute Meldrum's claim:

> I first visited the place on Sunday, June 12, 1892, accompanied by a black boy, and after satisfying myself that the remains were not those of either cattle or horses, but of some much larger animal, I secured a few specimens, which I brought in

> and showed to the men on the station, one of whom was Meldrum. On the following Sunday he, as before stated borrowed a horse from me, and, accompanied by the same black boy, drove out to the same place and brought in the bones, which he some four months afterwards exhibited in Adelaide. I intended visiting Adelaide early in October of 1892, and therefore did not forward my specimens by post, preferring on account of their fragile nature to keep them under my own care until such time as I could take them down. Unfortunately adverse circumstances prevented me leaving the station as early as I intended. I was totally unaware of Meldrum having taken his specimens to Adelaide and was first made cognizant of the fact by the appearance of a report in your paper.

The 'black boy' in this account played a background role, a nameless someone who 'accompanied' first Ragless then Meldrum, rather than led them to the bones.

Some time before the controversy reached the Adelaide papers, George Goyder, Surveyor General of South Australia, wrote to Stirling on 13 December 1892 informing him that a Mr F.B. Ragless of Enfield 'called at this office yesterday, and stated, that he is the discoverer of the fossil remains at Callabonna Station'. Ragless remarked that at that time he employed Meldrum and 'a black really was the discoverer of the bones & showed them to Mr Ragless on the following Sunday who subsequently showed them to Meldrum, also that there are from 12 to 14 specimens'. Goyder wrote again on 21 July 1893, as Ragless had enquired about his reward. Stirling in reply reassured Goyder that the museum regarded Ragless as the discoverer, but thought it premature to discuss rewards before the museum established the extent and value of the find. Neither Stirling nor Ragless at any time referred to a specific reward such as the £1000 purportedly offered by the museum and the Royal Society of South Australia in 1870.

Well into 1893, the Ragless brothers remained frustrated that the museum rewarded Meldrum for his 'duplicity' with £3 a week. Perhaps some of their later complaints against the Hurst party, and Henry Hurst's hints at the Ragless' recalcitrance, exorbitant charges and broken promises, sprang from resentment over Meldrum's presence in the Museum group, and the drought-threatened pastoralists' unease over the disposition of the much-disputed reward.[3]

On the other hand, no documents exist to prove the Ragless brothers' claim to the lake bed discovery with any more certainty than Meldrum's. Meldrum stuck to his story well after the Adelaide papers published the Ragless versions. A letter to the Board of Directors on 11 January 1894 enquired about his promised reward – an unspecified 'discoverer's claim'. He wrote in detail that 'during the year 1892' while engaged in well sinking, he received 'certain information ... from the Natives' which induced him 'to take one of them and explore the Lake for some distance'. Eventually they 'came across the remains of several large animals the bones of which I had never seen before'. He collected as many as he 'could conveniently carry and took them into the Station where they were seen by the men working there'. He 'kept them by' him for 'a couple of months' before bringing them to Adelaide in early November. His story is similar to Fred Ragless' much later account of his own interest in the giant bones, except that Meldrum's reference to learning from 'the Natives' has a generic quality, as if stories of the bones were a commonplace for the Aboriginal people at Callabonna Station.

Ragless' story of his introduction to the bones on the lakebed gained narrative colour over the years. In the first version which appeared in the *Register* in 1893, Ragless' older brother Richard explained that one day, an Aboriginal man told them about giant bones. Fred's subsequent accounts over the months and years became more detailed, culminating in the version in his memoir, written in 1936, of elephants and campfire stories, which introduced 'Jackie Nolan'. Perhaps Meldrum refined his story having read Ragless' account in the *Register*. But perhaps he told the truth. There is no particular reason that the 'Natives' or 'black boy' of their two versions should have told only one European man of the bones in the lake. Maybe Meldrum had already seen or heard of the bones when Ragless returned, fossil-laden, to the homestead, showing his new collection to the men, and was inspired to ride out himself. Maybe he had already collected the bones and Ragless' return merely provided the catalyst for his trip to Adelaide. Perhaps Meldrum acted in good faith, believing himself the discoverer of the bones. But if the Ragless' account is accurate, Meldrum inexplicably held onto the bones for four months before hurrying to Adelaide to claim palaeontological glory.

Certainly, he is Australian vertebrate palaeontology's main candidate for villainy, for all that he introduced the 'wilderness of fossils' to Adelaide scientific society. The descriptions by twentieth-century vertebrate palaeontologists of founding moments of the discipline in Australia regularly retell a version of the Callabonna story that credits Ragless with the discovery of the bone-yard, qualified with mention of Jackie Nolan's role in leading him there and John Meldrum's cupidity.[4] It suited the museum board better to regard Fred Ragless as the discoverer. After all, Ragless did not have a contract with the museum, with a clause vaguely promising a discoverer's fee.

'Incompetent to conduct a work of such magnitude'

The Hurst excavations

Henry Hurst's reports of the fossil field were initially promising. He requested that the museum seek to finance a much longer collecting expedition to the lake. Stirling put this to the cash-strapped board and eventually pastoralist, magnate and philanthropist Sir Thomas Elder came to the party and provided £500, which kept the museum team in the field until November 1893, albeit with a change of personnel.

Eager to be involved with the later reconstruction and analysis of the bones, Hurst wrote to Stirling of his determination that South Australia should have the honour of the discovery of the first complete Diprotodon skeleton, his assessment of the richness of the site, as well as the other megafaunal fossils the group found further afield in the lake. Stirling for his part complained of Hurst's irregularity in communicating with Adelaide. Hurst blamed the inclemency of the weather, which often prevented him leaving camp for Callabonna Station to send mail. H.Y.L. Brown visited the site in May 1893 and, although astonished by the lake's prehistoric wealth, wrote an excoriating confidential report to the board, scathing of the way Hurst ran both the camp and the excavation. He was appalled at the state of the exhumed fossils, which he saw exposed to the rain and wind and (where labelled at all) arranged idiosyncratically by bone type rather than individual specimen. Hurst's first trip back to Adelaide in July 1893 aroused Stirling's anxieties further. He had hoped for many more specimens and those that arrived were of a 'varied and miscellaneous character'. Many were much broken. Veiled and not-so-veiled criticism of Hurst began to appear in print amid concern that 'this important field of fossiliferous wealth would be conserved for the benefit of science and the colony'.[5]

Hurst blamed the transport by camel across a landscape alternately sandy and boggy, and the difficulty of protecting the fossils in the field. He wrote frequently, enthusiastically and in vain to Stirling requesting more camels to help in the transport of the specimens and in the procuring of rations and packing materials. He also mentioned the problems of the weather – either too hot and dry to keep horses near the camp or too wet for the excavation and packing to progress efficiently. He kept many specimens unpacked in long anticipation of Stirling's visit. Several times, he explained his decision to try to build a composite Diprotodon from the ground up, on the basis of the richness of the field and the probability eventually of unearthing a 'perfect' specimen.[6]

But it seems Hurst and Stirling had different collecting philosophies. Hurst believed that the most important part of his brief and the 'main object of the expedition' was to find or construct a complete Diprotodon, paying especial attention to evidence of the nature of its feet. By 11 March 1893, he wrote, he only required a few more bones 'to make up the total number'. Henry and his brother George were commercial collectors, freelancing when they had no institutional affiliation, for whom quantity was as important as quality, and more important than detailed provenance. Hurst reasoned to Stirling in the same letter that as he had not so far met with 'a skeleton which could be extracted absolutely perfect some portion or other being nearly always found missing', he considered it 'the wiser plan to get one by the quickest means possible'. They were operating to a tight schedule.

Hurst reasoned that the most efficient way to assemble a Diprotodon was piecemeal, 'by selecting the bones from a number of animals'. At any other site such a method 'would be an impossibility, but here the task is rendered easy through the remains being present in such fabulous numbers – thousands of skeletons are probably buried below.' They sought the 'best' parts from which the museum preparators could fashion a composite whole, waiting all the while to 'find one which is entirely complete in itself' when Hurst would 'take the greatest pains to get it out perfect – This I do not doubt I will do before we have finished'. It is clear from many of his letters that this is what he had set about doing, but Stirling seems to have missed or ignored the implications, for it was not until the first specimens arrived and more particularly after his own visit to the site that his ire was fully aroused:

> In place of having in view the great necessity for keeping together and distinguishing the bones of individual specimens, they were evidently indiscriminately mixed together. It was clear that Mr Hurst's unwise attempts to construct perfect skeletons from parts of different animals had led to inextricable mixing of the bones, not only of those of the same species, but probably of those of different species.[7]

Stirling at Callabonna

Stirling finally left Adelaide for Callabonna with Zietz and museum employee Thomas Cornock on the morning of Friday 11 August 1893, intending to leave the two museum officials in charge of the camp if all was not going to plan. They reached the camp on 17 August, only a day after Hurst, whose return from the city was hampered by a camel accident at Farina. Stirling stayed until the following Monday, 21 August, and reached Adelaide five days later. Zietz, Cornock and the rest of the museum party, minus the Hursts, returned to Adelaide the following December.

Stirling was critical of every aspect of Hurst's management of the camp and the excavation. He saw 'on all sides' evidence of 'utter carelessness, slovenliness, and want of system', as valuable bones lay about 'exposed in all directions, some fortunately still in fair condition, while others were irretrievably ruined by exposure and bad treatment'. Of the large number that had been labelled and stored under cover of the tents, Stirling found he could 'attach little value to the accuracy of the labels'. He found the camp 'slovenly in the extreme' with refuse, tools and utensils 'left indiscriminately littered about'. The interiors of the tents 'were disgracefully dirty and untidy – the original stock of cooking utensils were mostly broken or lost'. To cap things off, two bags of plaster of Paris 'had been allowed to be exposed to the rain and were entirely spoiled and although more than 20 boxes were available some four or five only had been packed up' with specimens. He found it 'difficult to imagine the Camp of a scientific Expedition in a more disgraceful condition'.

As far as the excavation itself went, he saw 'a want of method in the plan of operations', as bones which he was convinced 'might have been removed whole' had broken in the course of removal, and bones which were 'perhaps ... unavoidably broken' had not been kept together for subsequent mending. Most decisive for Stirling was the 'inextricable mixing of the bones', the result of Hurst's 'unwise' attempts to construct complete Diprotodons.

Stirling returned to Adelaide at the end of August, announcing the 'resignation' of the Hursts. He had left Zietz in charge with Cornock as his deputy. Disappointed and angry at the progress, he was nevertheless impressed with the fossil field. He wrote, in a confidential letter to the museum board, that 'careful and repeated examination' of the specimens Hurst brought from the lake precipitated the director's journey, having convinced him of three things. He believed that Hurst's method of preparation was 'in some cases actually prejudicial' to conserving the bones; he claimed Hurst's organisation and labelling of the bones to be unreliable; and his reports were 'unsatisfactory and incomplete'. In addition, Brown's earlier scathing report claimed that 'the work of excavation was not being wisely conducted'. On the other hand, Hurst's accounts 'incomplete as they were', as well as that of Brown, 'could leave no doubt of the extraordinary richness of the deposit in fossil remains of great variety'.

Not that Stirling was entirely oblivious to the difficulties presented by Lake Mulligan–Callabonna as excavation site (see Figure 24). Aside from its undeniable remoteness from sources of provisions and the difficulties of obtaining enough of the invaluable camels, there was the weather, of which Hurst had complained in several letters to the director. It did not take long to make itself known to Stirling. On the evening of 18 August, the day after his arrival, 'rain began and continued all night, so that by next morning the bones, though covered up as well as we could, were too damp to be packed away'. In addition, the 'tenacious clay of the camping ground became also horribly soft and sticky and it was in fact impossible to get satisfactorily on with anything'. So he concluded the body of his attack on Hurst's management with a small proviso:

> Whilst therefore I have animadverted somewhat severely upon Mr Hurst's method of management, I fully recognize that in the face of existing difficulties the work of keeping the camp going is, alone, a matter requiring a very considerable amount of management and foresight, to say nothing of the difficulties in dealing with the bones themselves, which require the most careful treatment.

He did not soften his assessment of Hurst's performance with this qualification, claiming that the collector had throughout 'shown himself to be completely incompetent to conduct a work of such magnitude'. Under Hurst's management 'a fraction only of the success which was possible, has been achieved'. Although Stirling acknowledged that the team under Hurst excavated 'a considerable quantity of material, probably when packed up amounting to three tons', he lamented that 'Mr Hurst has not yet got a single complete individual skeleton of Diprotodon' and Stirling thought it 'exceedingly unlikely even that a single skeleton can be made up'.

The Zietz excavations

Stirling continued to underestimate the difficulties involved in preserving Callabonna specimens, which remains a problem at the site. They are impregnated with gypsum (hydrated calcium sulphate), which crystallises when the fossils dry, causing the bones to expand until they explode and fragment. But the site itself, while not redeeming the Hurst brothers, went some way to compensating the museum for the time and money already expended. Stirling

Figure 24: View of Lake Callabonna, 1893. (South Australian Museum expedition, 1893. In *E.C. Stirling 1900.* The physical features of Lake Callabonna, *Memoirs of the Royal Society of South Australia,* 1(2): 1–15. *National Library of Australia Nq 560.9942 ROY (2053869). Stirling's explanation: View, looking south-west, showing part of the flat saline expanse of Lake Callabonna, with the western shore just visible as a dark streak. The elevation in the foreground is the top of the sand-dune at the foot of which, on the further side, somewhat to the right of the erect figure, the camp was situated. The vegetation is stubbly samphire (*Salicornia*). The bulk of the fossils were obtained on the flat to the (observer's) right of the erect figure on the sand-dune.)*

told the board that 'in promise of results, the field far surpasses any expectations which previous reports had led me to form of it'. He anticipated that there were 'hundreds or even thousands of skeletons yet untouched and the ground actually explored is but a minute fraction of the whole fossil bearing area' (see Figures 21 and 25).

Stirling described the successful method used by the Hursts (without crediting them) to find subterranean bones, probing with a wire rod, whereby he 'struck underground skeletons or parts of skeletons about a dozen times in a very short space of time', and he expected that the conduct of subsequent operations, under the methodical and careful Mr Zietz, would 'yield results which in scientific value will far exceed those already obtained'. Accordingly, he 'told Mr Hurst [his] opinion' and 'gave him and his brother ... a week's notice in terms of the last agreement signed by him, and desired that the third brother (Wm. Hurst), who had been a sort of hanger on of the camp for the last six weeks, should also leave'. Henry Hurst asked Stirling 'that he might be allowed to send in his resignation', to which the director assented.

The museum committee held a special meeting on 31 August to discuss Stirling's visit, at which he summarised his report, which itself borrowed heavily from his expedition journal. He informed them that he had left Zietz in charge with Cornock as his deputy. He retained Meldrum at a reduced rate of pay, and the other excavator Templeton, both of whom he found 'willing

Figure 25: Diprotodon fossil in Lake Callabonna showing surface expression. *(South Australian Museum expedition, 1893.* In *E.C. Stirling 1900.* The physical features of Lake Callabonna, *Memoirs of the Royal Society of South Australia,* 1(2): 1–15. *National Library of Australia Nq 560.9942 ROY (2053869))*

obliging and handy'. Stirling regretted 'the stringency of the measures I have been compelled to take, in the interests of the Museum' and he was disappointed 'in regard to the inadequacy of the results so far achieved the great expenditure into which we have been led by a continued course of mismanagement'. But he assured the board that 'failure in the future to make the best of this extremely important discovery' would compound the sin and 'be humiliating in the extreme to all concerned'. He hoped that 'the value of the future work will redeem the comparative failure of the past efforts to make the best use of a field of extraordinary promise'. He saw no difficulty that could not be surmounted 'with patience, judgment, and foresight, save and except a continuance of wet weather which must completely paralyse all operations'.

As recounted by Stirling and Zietz in 1899, the party under Zietz' command remained at the lake until November 1893, at which point they were indeed driven away by rain, having previously been wracked by drought, rabbit plagues, trachoma and gastric illness caused by dead rabbits in the water supply. The Ragless brothers lasted five years more, but their stock was decimated by drought and dingoes, which had reached plague proportions as the rabbits increased.[8]

Contested possibilities: in defence of Henry Hurst

Unsurprisingly, the Hursts did not share Stirling's view of recent history. A half-hearted effort to write Henry out of the story appeared in the *Register* of 22 March 1894. George Hurst replied on 7 April in a letter to the editor, expressing regret 'that no reference was made to the earlier party in charge of my brother'. However, no hint appeared publicly to suggest anything remiss about the Hurst brothers' 'resignation' or their performance at the lake, aside from an earlier and anonymous letter in the *Register* in June 1893, presumably from Brown, decrying the state of the excavation and criticising Hurst's management. Otherwise, the museum board kept its concerns and recriminations mostly private.

In 1918 Hurst, then based in Brisbane, wrote to Stirling (who had resigned the directorship in 1913), explaining that he had always felt the director treated him unfairly. In the 25 years since his dismissal, he had visited Zietz more than once and established that 'very little was done after I left the Lake and scarcely any (if at all) specimens were obtained'. Zietz attributed this to 'the specimens crumbling on exposure to the air'. Hurst claimed never to have experienced such a difficulty (unlike most subsequent excavators at Callabonna) owing to the adoption of 'special measures', which he 'again put successfully into practice when on similar work in Europe'.

Indeed Hurst had by his own report spent much time away from Australia, and initially, away from palaeontology. Now the well-travelled and apparently rejuvenated Mr Hurst was 'very keen on making fresh discoveries of Post-Pliocene vertebrates'. His eye turned once more to the Lake Callabonna region as he was 'in a position to locate at once some valuable finds (from information obtained lately)', certain of 'a successful termination'. Stirling forwarded the letter to Edgar Waite, his successor, who thanked Hurst warmly, but had to inform him that 'there is, at present, no intention on the part of this Museum of further exploring the Lake Callabonna District'.[9]

By 1893 Hurst was an experienced collector and palaeontologist, but he had no experience leading an expedition of the size and logistical requirements of the Lake Mulligan site. His camp and site management might well have been inadequate. But there were many mitigating circumstances beyond his inexperience, as even Stirling recognised though he did not acknowledge it. At times Hurst showed considerable innovation. Aside from his brothers, the men on his team had no experience as excavators, and Hurst actually showed some initiative in training them. For example, when the expedition first reached the lake, they began digging at Meldrum's site, but the specimens there proved to be 'much decayed'. So instead of attempting to recover them for the museum, Hurst used them to train his inexperienced workers. Brown's visit to the site on 25 May 1893 lasted less than a day and followed several days of light rain which troubled the excavators. The government geologist was hardly in a position after such a short visit to assess the circumstances peculiar to Lake Mulligan which might have affected the diggings. As far as Stirling's complaints about the piecemeal attempt to construct a Diprotodon from the parts of many different specimens, Hurst had kept his employer informed of his plans and technique. By ignoring this, Stirling gave tacit agreement to his methods. Hurst requested Stirling's presence at the site many times.

Having been at the lake for five months with essentially no guidance, other than the little Stirling provided sketchily in several letters, which took up to a fortnight to arrive, Hurst was hastily summoned to Adelaide, where the museum committee informed him that Stirling and Zietz planned to follow him back. Unhappy with the specimens he brought to Adelaide and with their condition, the board was nonetheless unable to provide him with more camels or packing crates. He had left the camp in charge of his brother George and his return after two full months, during which he was either in Adelaide or in transit, was delayed by a camel accident. He was thus unable to tidy the camp in preparation for Stirling's long-anticipated arrival. Brown and Stirling both complained about exposed specimens at the camp, yet Hurst had written many times to Stirling telling him these were left unpacked 'in anticipation' of the director's visit, for the purpose of showing them to him.

Clearly Stirling was embarrassed at the progress of excavation at the museum's showpiece. The expedition had absorbed a large amount of public money, and the Board of Directors was now answerable to Elder as well. The museum needed a good return, with hope of financial as well as intellectual benefits. Adelaide hosted the fifth meeting of the Australasian Association for the Advancement of Science in September 1893. Ralph Tate, with Stirling one of the museum committee members responsible for Hurst's appointment, was president elect. Stirling and Tate had hoped to have three and perhaps four complete skeletons ready to display at the meeting. The committee needed scapegoats, which Hurst's somewhat unorthodox methods in a difficult environment provided. Furthermore, from 1889 to 1894, Stirling was honorary director of the museum. At the end of 1894 the board elected to pay him a full salary. The anticipation or hope of this might have influenced his behaviour. It was important to him personally as well as for the prestige of the institution he led that the investigation of Lake Mulligan–Callabonna proceed smoothly and successfully.

In his enthusiasm for the site and his eagerness to continue excavating, Hurst somewhat naively wrote to Stirling in March telling him that 'The material we now have is worth many hundreds of pounds at the least and I do not think we have got a fractional part of what another 3 months work would yield if we continued the search.' His announcements in April and May that his party had unearthed 'three nearly complete skeletons' together with 'over 2000 separate bones of the same animal belonging to over seventy individuals' raised Stirling's expectations that the museum might assemble a complete specimen before the AAAS meeting. The director's disappointment was therefore proportionate to Hurst's promises when the expedition leader arrived in Adelaide with far fewer specimens than anticipated and in poor condition. As Hurst later wrote, Stirling did not, at the time, fully understand 'the peculiar difficulties involved in safely unearthing the Diprotodon, and other bones'. He subsequently acknowledged in a number of articles in the *Transactions of the Royal Society of South Australia* and in *Nature* that conditions at the lake compromised excavators' ability to capitalise on its fossil wealth, but did not vindicate Hurst.[10]

The recuperative power of the deep past

The *Register* rapidly adopted a cabbalistic voice when the first Lake Mulligan bones reached Adelaide, celebrating the region's presumed former lushness, as revealed by its mysterious material remains. In August 1893 the newspaper linked the 'sterile' landscape to the wonder of the deep past: 'the gigantic creatures of a bygone age whose relics are embedded in the hardened mud and débris of what were once the huge swamps of the Far North-East'.[11]

Thirteen months later, another anonymous scribe in the *Register* described the Callabonna site, located in 'a series of 'claypans' in the vast desert north-east of Farina', as 'a 'lake' only in flood times – an arid waterless region of sand and salt pools … where even rabbits do not thrive and camels are a necessity'. Ornamenting this scene are 'the bones of monsters' which 'lie as they died' (see Figures 25, 26). However, far from contributing to the Hadean character of the site, the bones raised questions that began to recuperate the reputation of the lake environment:

> On what did they feed? Why were they made so enormously strong? … What astonishing changes in the climate and rainfall must have occurred since these beasts roamed in the jungle as the rhinoceros, or in swamps as the hippopotamus? … These things are among the buried secrets of the great past, which secrets science is unlocking.[12]

Figure 26: 'The bones of monsters which lie as they died'. *Partly excavated Diprotodon fossil in Lake Callabonna, South Australian Museum expedition, 1893.* (In *E.C. Stirling 1900.* The physical features of Lake Collabonna, *Memoirs of the Royal Society of South Australia,* 1(2): 1–15. *National Library of Australia Nq 560.9942 ROY (2053869). Stirling's explanation: Skeleton of* Diprotodon australis *in process of excavation. The head appears in the right foreground; the right humerus is extended to the (observer's) left; a stick crosses, and is bound to, the pelvis, and the tail is seen resting on a white patch)*

Again in the *Register*, in May 1893, the bleakness of Callabonna, 'a more apparently uninteresting and forsaken spot' than which 'could not be found in the whole of Australia', is juxtaposed with a fulsome account of its glorious deep past, as revealed by science:

> A different picture of the past history of this country is brought to light by the new discoveries! [Nearby] Mount Searle [sic] ... was probably at the period of which we speak fully twice [its present] elevation. On its sides grew huge trees, and all around was a dense tropical growth, exceeding in luxuriance the forests of the eastern slopes of the Andes in South America. Every form of life on the whole earth was at that time huge and uncouth ... [These skeletons] are a magnificent treasure in the interests of science ... Of very far-reaching significance is the story which is told by these mute relics of a remote age.[13]

This discovery 'in the most unexpected' location, 'of incalculable value to science', went some way to redeeming the lake as a functional colonial landscape. At first glance, the story reads as one of degeneration from a Pleistocene Eden to a modern desert, paralleling modern geological narratives and the *kadimakara* accounts. Closer reading shows that value in the modern landscape is clearly attached to the wealth of meaning, and potential real economic wealth, the bones provided.

In an article in *Nature* in 1894 and again in 1907, Stirling dismissed the scenic merits of the lake in unflattering terms as 'Almost unsurpassable for barrenness and utter desolation'. He made reference to names 'such as Mount Hopeless, Dreary Point, Illusion Plains, Mount

Deception, Mirage Creek' which invoked the 'character of the surrounding country' in which 'Scarcely any vegetation relieves the prevailing desolation.' As for the lake itself, 'not a bush relieves the unbroken monotony of the level, white crystalline surface'. But he considerably tempered his criticism of the region's sparse severity with the statement that 'There is compensation for the uncompromising physical features' of the lake 'in the fact that its bed has lately been shown to be a veritable necropolis of gigantic extinct Marsupials and Birds, which have apparently died where they lie, literally in hundreds.' The Adelaide press equated the economic promise of the fossil site with the lucrative copper finds in the colony earlier in the century and even at the height of his disappointment with the conduct of the excavation under Hurst, Stirling was convinced that the richness of the lake would eventually compensate the museum fully, in scientific if not economic terms. The site's scientific promise was rooted in Diprotodon feet, tails and epipubic bones.[14]

Inspired by the emerging stories of Callabonna Station, Mr James Ward wrote to the *Register* in July 1893 describing the state of the north of the colony 'in earlier times', as told to him 'by a blackfellow at a marshy swamp on the Wilpanara Run in the year 1865'. The Aboriginal man described 'some huge animal that at one time did exist' and which dwarfed a horse, as recalled to him by an older relative. Ward 'was impressed that if such an animal ever existed, it only frequented marshy places'. Based on his experiences in the country north of Adelaide, and the advice of the Aboriginal man, he was in no doubt that Lake Mulligan–Callabonna 'has been a large marshy swamp at no greatly distant period'. In support, Ward recalled that 'in or about the year 1855', land 'between the Para River and Virginia was rich and fertile, bearing heavy crops', with 'paddocks of wheat three feet high, and in the black sandy hollows over four feet'. By 1880, the land was comparatively barren 'and the hollows salt lagoons'. Wells in the district sunk by the early residents 'were from forty to fifty feet to the water. In many of them it has risen to within a few feet of the surface and become quite brackish.' Based on the rapidity of change in the landscape, he hypothesised that 'If the sea encroach inland at the head of Spencer's Gulf at the same rate it is probable that in a thousand years hence ships will sail to Lake Torrens.'[15]

Rather than 'the sea encroach[ing]', this letter appears to represent an early account of the salinisation of South Australia, with a rising water table and encroaching salt following the extensive land clearance and replacement of native flora with shallow-rooted European crops that accompanied European expansion into the interior of the colony. Ignorant of longer cycles of humidity, aridity, groundwater fluctuations, and of Aboriginal husbandry of the inland, the Callabonna remains evoked a fertile recent past and the promise eventually of a navigable inland sea for James Ward, just as explorers like J.W. Gregory saw the potential recovery of the 'dead heart' in its well-watered deep past and Charles Sturt sought salvation in an inland sea. Contributors to the *Register* and *Observer* inferred the former existence of a world where life was bigger and lusher than modern Australia, but the articles hinted at possibilities for contemporary exploitation as well, in reference to ivory, mineral wealth, knowledge and prestige. Once liberated from the 'wilderness' of Mulligan–Callabonna, the fossils gain meaning and in turn imbue the lake landscape with that promise. Thus fossils replenish and recuperate country that is in other contexts deemed useless or barren.

South Australian parochialism, intellectual capital and the mobilisation of the deep past

Richard Owen, international bone collector, named *Diprotodon optatum* in 1838. 'Optatum' means wish or desire, and in 1877 Owen recorded these words that could be a palaeontologist's manifesto: 'Of no extinct animal of which a passing glimpse, as it were, had thus been caught, did I ever feel more eager to acquire fuller knowledge than this huge marsupial. No chase can equal the excitement of that in which, bit by bit, and year after year, one captures the elements of reconstructing the entire creature of which a single tooth or fragment of bone may have initiated the quest.'[16] By the time he died (by now arguably Europe's leading comparative anatomist), most parts of the animal's skeleton had been found and sent to him for identification, but no foot bones had been identified. The anatomist could speculate on Diprotodon locomotion, but much about its stance remained a mystery. His desideratum was the foot (see Figure 23, page 98).

In November 1892 Stirling announced the discovery of an abundance of Diprotodon fossils at Mulligan's Lake. In December of the same year Owen died, aged 88. The year after Owen's death, a cast of a Diprotodon foot was discovered in the sediments of Mulligan's Lake. That same year the first foot bones were found, putting to rest any last doubts as to whether it had feet. H.Y.L. Brown returned from the north of the colony in early June, having spent one day at Mulligan–Callabonna. He reported in the *Register* on 13 June 1893 that 'parts of 80 Diprotodon skeletons have been found, including feet and tail, and an impression of the foot with five claws' (see Figure 27). The discovery of its tail bones promised to identify its closest living marsupial relations. Diprotodon foot-bones had in fact been discovered many years before at the Wellington Caves and sent to Owen, but he or his correspondents assigned them to other skeletons. When Krefft deduced in the 1870s that some of the mysterious metatarsals and phalanges must belong to Diprotodon, his suggestions were entirely ignored.[17] In any case, how could such a delicate toe belong to this marsupial behemoth? The great advantage of the Callabonna remains, aside from their extent and variety, was that they were found *in situ*, their articulated legs and feet leaving little doubt as to which feet belonged to which legs (shown in Figure 27).

Owen died without recognising the feet right under his nose. But their significance surpasses the narrative twists their discovery invites. Like minerals, fossils were part of the capital of the earth and the wealth of empire. Colonial authorities saw an opportunity to exploit the new resource, as witnessed by the promise of cash rewards. When the museum and the Adelaide newspapers announced the wealth of the Lake Callabonna finds in 1893, metropolitan scientists anticipated a marsupial windfall. It was not forthcoming. Even during the great palaeontologist's lifetime, his control of the lost worlds of reptilian, avian and mammalian megafauna began to slip, particularly as the astonishing collections of dinosaur bones from America's Midwest became apparent. Following his death, the British Museum of Natural History's hold over Owen's empire of giants crumbled further. Stirling held onto his feet.

Mechanistic models invoking a unidirectional flow of data unmediated by ideology or analysis from the colonial 'periphery' to the metropolitan 'centre', and of knowledge and patronage back from the centre to the periphery have characterised colonial collectors, naturalists and institutions, particularly during the first three-quarters of the nineteenth century, as handmaidens to imperial science practised in Britain. In particular, teleological assumptions frequently underlie vertebrate palaeontologists' accounts of the history of their discipline in Australia and its development from a pioneering age of discovery to the future with its 'unfathomed areas in need of exploration'. But metropolitan theories and classifications were not formulated in a vacuum. Ditch-diggers, sheep farmers, Aboriginal people, collectors, amateur naturalists and increasingly professional scientists selected which parts of the fossil record made their way to which scientist, in a chain of serendipity, patronage, decreasing agency and increasing renown, beginning with the coincidence of preservation and discovery.

The timing and extent of the singular Callabonna discovery was fortuitous for the South Australian Museum. The 1893 AAAS meeting was a boon for Adelaide's cultural institutions generally. Although the museum's anatomists, articulators and taxidermists did not have time to assemble a complete Diprotodon before the meeting and Stirling's paper on Diprotodon feet

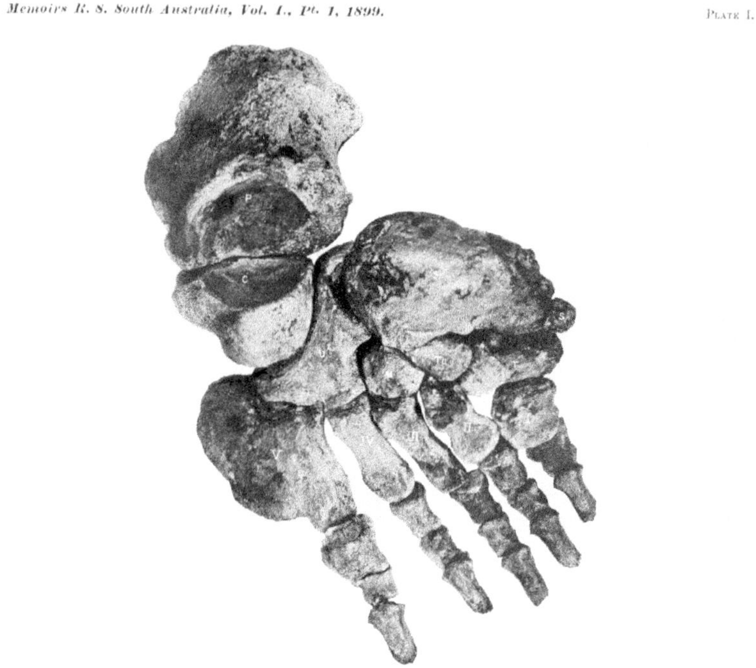

Figure 27: 'Year after year, one captures the elements of reconstructing the entire creature of which a single tooth or fragment of bone may have initiated the quest.' *(Articulated right manus [Diprotodon]. Plate 1. In E.C. Stirling and A.H.C. Zietz 1899b.* Fossil remains of Lake Collabonna, Part I: Description of the manus and pes of *Diprotodon australis*, Owen, Memoirs of the Royal Society of South Australia, 1 (1): 1–40. *National Library of Australia, Nq 560.9942 ROY (2053869))*

did not appear for another three years, both the meeting and the fossil remains generated a great deal of popular interest, feeding off each other and the consequent media flurry.

Between November 1892 and September 1894, close to 60 items about the Lake Mulligan megafauna appeared in the *Register*, often with duplicates in the *Observer* and *Advertiser*. More than 40 of these appeared in 1893. In September 1893, Ralph Tate delivered a lecture to the Royal Society of South Australia on the evidence for climate change offered by the Callabonna remains. Also in September Charles De Vis' lecture to the AAAS on 'the Diprotodon and his times' received much favourable media coverage. There was also popular interest in the fossils overseas. The British Association for the Advancement of Science reported on the remains.[18] On 25 April *The Times* published a telegram, subsequently published in *Nature*, from Stirling to Alfred Newton of Magdalene College, Cambridge, describing the exciting discovery of Diprotodon foot-bones. Reports in the *Scotsman* appeared in late 1893.

At first glance the Diprotodon's toe may seem to have little to do with South Australian parochialism and the commodification of the material remains of the deep past. But as a letter of 11 March 1893 to Stirling from Henry Hurst while the latter was stationed at Callabonna, made clear, 'the main object of the Expedition' as Hurst understood it was that South Australia would 'have the *honour* of possessing the first complete Diprotodon'. An increasing demand for independence, recognition and specialisation among colonial museum curators and naturalists during the last quarter of the nineteenth century resulted in more aggressive collecting and exchange policies in colonial institutions, epitomised by Gerard Krefft's lament at the inadequacies of metropolitan classifications of Australian fossil material.

Specimen exchange aided the accumulation of scientific capital. Museums attached considerable cachet to the acquisition of exotic foreign materials. Unearthing duplicate specimens which could be used for exchange with other institutions facilitated this, as in the Callabonna case, and the construction of casts, as museums were loath to risk the loss of unusual or valuable specimens. As Hurst wrote to Stirling in excitement from the field, requesting more funding to prolong the expedition in light of the unexpected concentration of fossils, 'By the end of the period ... I believe there would be enough material to supply a large number of the Museums in the World.' 'Private interests' speculated in 1893 that the Callabonna fossils would be more lucrative than copper mining or ivory and in 1894 an Adelaide syndicate was floated with the intention of unearthing Callabonna specimens for private profit. Henry Hurst at the very least would have approved of such a proposal, continuing to Stirling in his letter of 16 March that 'It would be a great pity to suspend operations so hastily as it is now proved ... that there is a great possibility of more valuable discoveries.'

Stirling and the museum guarded their intellectual capital. For example, in his 1894 *Nature* article, Stirling wrote cagily that 'Of the feet, in which from our previous ignorance of their constitution much interest is centred, I prefer not saying much at present.' Contrary to mid-century trends, the director and Amandus Zietz described most of the specimens themselves and published their most significant findings in the *Transactions* and *Memoirs* of the Royal Society of South Australia before more prestigious overseas journals, despite the interest from abroad as well as within the colony. Apart from Stirling's two accounts in *Nature*, between 1894 and 1914 Stirling's and Zietz' papers on the fossil remains of Lake Callabonna appeared in Royal Society of South Australian publications and South Australian Museum memoirs. The director met correspondence from the British Museum requesting samples with firm apologies. He explained that it was impossible to transport the skulls without risking irreparable damage and many of the other fossils were not fit to travel. To prove his goodwill and the patriotism of the museum board he was willing to trade some metatarsals and maybe a clavicle. But in the interests of science and all their institutions, he advised waiting until the museum had got 'all the parts accurately represented' and a cast of the whole might be available.[19]

The board was reluctant to allow scientists from other states and institutions untrammelled access to the site, as discovered by J.W. Gregory when he plotted his demented summer field trip in 1901–02.[20] In September 1901, during the planning stages of the expedition, Gregory wrote to Zietz (as Stirling was in England) that as he had discovered that Callabonna 'was a proclaimed reserve of your Museum' he had 'thought better' of making it the main feature of the field trip, which had been his initial intention. But he was still eager to collect at the famous site and asked Zietz accordingly if there would be 'any objection to my going ... for a week or two's collecting' during the return trek from Lake Eyre to Melbourne. He hoped during his time in the Lake Eyre basin 'to collect, or try to collect some bones, and get a general idea of the structure of that country' where the Pleistocene remains were found in such great variety and number. He also looked to compare the desert conditions with those of Africa, where he had found bone deposits forming around dried waterholes, in a clear analogy with Callabonna.[21]

Zietz wrote to the museum board that he did 'not think that he could hurt us, if he should carry out his intended visit'. Nonetheless, protective of the museum's intellectual interests, he advised 'that Prof. Gregory be informed that owing to Prof. Stirling's absence you would not like to take any steps in his favour' and thus 'it would be necessary to wait for his return about the middle of November'. Gregory replied, thanked Zietz for the information and suggested that he would only go to Callabonna 'for purposes of comparison' with Lake Eyre, which he found much more interesting, but should be glad of permission from the museum trustees to visit and collect at the site all the same. He tried carefully to reassure the board that of course he 'would not do anything to forestall the work of the SA Museum in publication over Callabonna'.[22] The trustees gave only grudging permission to the Melbourne expedition to collect at Callabonna, although the museum had not sent a collecting party there itself for eight years. Gregory wrote to thank them, but impatient with the museum's jealous provincialism, he added that as the Lake Eyre basin, 'to me the most interesting area in Central Australia', held quite enough attractions, he would most likely return to the west and bypass Callabonna altogether, as indeed he did.[23]

It was for these reasons of intellectual ownership and the control of meaning, as much as to protect the site from amateur rock hounds who might damage it and entrepreneurs who might exploit it, that the fossil reserve was first proposed by H.Y.L. Brown in 1893. He declared that the site afforded

> [scientific] opportunities which have seldom before occurred on this continent, and which may not occur again. The work now being done by the Adelaide Museum in collecting specimens will doubtless lead to the settlement of many points ... In view of the importance of preserving these relics of a bygone age for the future scientific exploration, I would recommend that the whole area ... be reserved for that purpose, and to prevent the indiscriminate digging up and removal of portions of the specimens.[24]

Requests for Callabonna fossils were made by institutions as far apart as Perth, London, Cape Town and Brazil, as well as the other Australian state museums. Eventually the museum distributed seven casts of the composite Diprotodon skeleton along with bones of extinct kangaroos, giant wombats and *Genyornis*, the giant flightless bird first identified from Callabonna remains, in exchange for such exotica as black monkeys (*Colobus guereza*), mummies, stuffed 'mammalian heads' and leg bones of the dinosaur Diplodocus.[25]

The popular significance of Lake Callabonna emerges primarily from the whisper of former worlds provided by its dirt and fossils, material remains crucial to the reconstruction of the deep past in South Australia. As information was literally and serendipitously unearthed in colonial Australia, cultural institutions began to build up indigenous collections, retaining duplicates for trade with other institutions. The roles of patron and client were further destabilised.

Stories of the material remains of Lake Mulligan–Callabonna's former worlds and their intellectual heritage continue to shed light on the perception of fossils as a mineral commodity, and on the importance of intellectual capital in late colonial museums in Australia. These stories intrude into the socio-political history of the colony/state. In 1893 (and again in 1953–54 and 1970) scientists, journalists, legislators and their audiences mobilised conceptions of the deep past, its material remains and intellectual legacy to insert Mulligan–Callabonna into their collective 'proto-national' or parochial identity as a place worth saving. In turn, the prehistory represented by the lake landscapes and material remains connected (and continues to connect) South Australia's (and South Australians') lineage with a nebulous, global deep past.

Losing Lake Callabonna

Before the revelation of its palaeontological wealth, Lake Mulligan–Callabonna was chiefly renowned as the easternmost shore of Eyre's immense, imaginary horseshoe lake of impassable salt flats and mud. Babbage, P.E. Warburton and A.C. Gregory atomised the great 'Lake Torrens' in 1858 and Lake Mulligan entered pastoral and cartographic consciousness. Its principal attraction lay in its distance from the wheat frontier. Notwithstanding Fred Ragless' evocation of 'blazing parakela' and 'flowering winter herbage', Gregory's tag – 'sterile and of little practical value' – stuck (Figure 28).[26]

When John Meldrum brought his collection of bones to Adelaide in 1892, the Adelaide press and scientific community rapidly incorporated the site into a canon of significant places. Stirling and Zietz published their last papers on the Callabonna remains 20 years later in 1913, as parts 4 and 5 of the *Memoirs of the Royal Society of South Australia*. Stirling died in 1919 and his obituary in the society's *Transactions* devoted some detail to his work on Callabonna. The museum continued to accord limited celebrity to the composite Diprotodon skeleton, whose cast for many years graced the entrance room of the Australian wing of the South Australian Museum until displaced by a whale skeleton. Since 2001 it has been partially reinstated in the Australian Fossil Gallery of the refurbished museum, and the National Museum of Australia in Canberra also has a cast.[27]

The museum excavation team left the lake in November 1893 and never returned. Lake Callabonna itself lay largely unconsidered by palaeontologists for 60 years. In September and October of 1916, Edgar Waite, Stirling's successor as director of the South Australian Museum, led an expedition to the Strzelecki and Cooper Creeks, taking advantage of the collecting opportunities presented by the drought of 1913–15. Although the lake was not part of his collecting itinerary, he passed close by, leaving a damningly brief record of the 'encrusted and barren area', evocative of Eyre's, Sturt's and Augustus Gregory's earlier descriptions of an inhospitable, charmless, useless landscape. Waite declared his passage to Mt Hopeless Station 'the most dreary country it has been my lot to traverse', reflecting that it is no wonder Eyre 'bestowed such a name – no tree, no bush, no water; and we had difficulty in scraping up sufficient material for fire to boil the billy'. The following day saw no relief: the party 'reached a waterhole and later the crossing between Lakes Blanche and Callabonna, all dry'. In 1918 Waite informed Henry Hurst that the museum had no further interest in the Lake Callabonna site.[28]

Figure 28: The remains of Callabonna Homestead. *(Photograph John Magee)*

This second 'loss' was largely determined by climate and economics – somehow different from the indifference, misunderstanding, misrepresentation and neglect of the earlier nineteenth century, which was driven by dashed hopes and pessimistic misconceptions. In 1928, a quarter of a century after the museum team under Zietz abandoned the site to dead rabbits and sand, Robert Bedford of South Australia's privately operated Kyancutta Museum visited the area seeking Diprotodon remains. During the 1940s a team from the South Australian Museum attempted to reach the fossil site but, hindered by weather, did not reach it. In 1945, Harold Fletcher went 'West of the Darling' in search of fossils, accompanied by fossil collectors George Alder and Frank Foster (respectively a tank sinker from Broken Hill and a government livestock inspector stationed at Tibooburra), and picked up a Diprotodon jaw on the lake floor, although they did not visit the Hurst/Zietz site of 1893. Once 'discovered' in 1892, the idea of the lake continued to entice and frustrate, with its 'aura of mystery' as Fletcher called it, the half-promise in the desolate countryside of an inaccessible, fragile, unbelievably detailed glance towards Australia's more verdant deep past.

Chapter 7

FINDING LAKE CALLABONNA AGAIN, 1913–1970

'Splendid prospects'

Six decades after the South Australian Museum packed up its fossil menagerie, the lake was a successful port of call for Ruben Stirton and his student Dick Tedford on their first trip to Australia in 1953. Stirton, whose name became inextricably linked to Callabonna in the annals of South Australian science, is credited with inspiring a renaissance in the study of vertebrate palaeontology in Australia in the 1950s. For some time he had been actively seeking funding for

> an idea which I have had in the back of my mind ... the possibility of finding fossil mammalian remains on continental Australia that would throw some light on the origin of the Monotremes, and their relationship to the mammal-like reptiles ... I have had unusual success in Colombia with a similar project in cooperation with their Geological Survey ... and if we do find fossil remains of early Tertiary Monotremes and Marsupials, it should be one of the most important discoveries that has ever been made in mammalian paleontology.[1]

Advised by Sir Douglas Mawson that 'we should find Tertiary mammals east of Lake Eyre if they could be found anywhere on the continent', Stirton decided to begin with the Lake Eyre, Strzelecki and Tirari Desert regions.

He followed in the turn-of-century camel-prints of Brown, who surveyed the Lake Eyre region in 1892–93 and 1894, and of J.W. Gregory's whirlwind palaeontological field trip in the summer of 1901–02. Neither expedition returned with identifiably Tertiary mammal remains but, based on a combination of his Colombian experience and contemporary geological commonsense, Stirton believed that the former lakes and creek beds of the region showed promise. As he wrote to one of a number of potential funding bodies the following year in the hope of continuing the success of 1953, Australia had 'been a large continental mass for so long' that it was 'only reasonable to assume that somewhere in the interior we should find badlands of continental deposits (river, lake, and floodplain formations)' with fossil bones of mammals.[2]

On the 1953 excursion, Stirton and colleagues also travelled to the Menindee Lakes on the Darling River to investigate the South Australian Museum ethnologist and polymath Norman Tindale's discovery of extinct megafaunal remains in possible conjunction with human artefacts and burials, and to Lake Callabonna, both trips at the request of the museum, and in exchange for labour and a truck.[3] But although the two Quaternary sites offered 'splendid prospects', Stirton was primarily concerned to disclose some of the little-known marsupial assemblages of Tertiary Australia. The more widespread and studied Quaternary remains of Callabonna and Menindee held less intellectual interest for the Texan palaeontologist, who dreamed of trying his luck 'at opening the door to a new field of research and to unravel a history of life that scientists have speculated on for a century':[4]

> It took considerable time and much discouraging prospecting to find the key to our problem, [despite] in the interim [discovering] the presence of gigantic extinct marsupials in human cultural levels [at the Menindee Lakes], and [rediscovering] the famous *Diprotodon* (marsupial the size of a rhinoceros) locality at Lake Callabonna ... [Our] later explorations along the Cooper's Creek east of lake Eyre were of much greater significance.[5]

Before returning from Australia in 1953 he wrote to the eminent Berkeley palaeontologist Charles L. Camp that 'the last trip of course is far more important than the other two'. In a similar vein if somewhat more bluntly, Dick Tedford later wrote that, notwithstanding the success and interest of the Callabonna and Menindee expeditions, their nine-month stay as Fulbright scholars 'had come within ten days of being a complete failure', before they found Tertiary mammal fossils in abundance.[6]

'We hit bone'

Despite the suggestion that the Menindee and Callabonna localities provided something of a consolation prize, Stirton's excitement is palpable in the accounts of his discovery of the original fossil locality at Callabonna, untouched by scientists for nearly 60 years and still incredibly rich in remains. When he first set eyes on the lake in May 1953, and rediscovered Hurst's and Zietz' 1893 camp, the desolation of the surface barely raised a comment. The nonchalant explorer–geologist with an eye for detail was making science:

> I finally wandered back on the lake floor and came into the high dune ... I could see saline flats island-ward from the dune before I reached it ... Off to my left I saw an object that looked like a box-board. So I strolled over that way ... I then came upon a kangaroo skeleton that commanded my attention for a few minutes as I looked for the cranium and mandibles. We have been picking up the best ones ... Next I saw some forked posts about half buried that seemed to form the uprights for a shelter. All this and other objects about rather convinced me that I had found Zeitz' [sic] camp site. Then I thought, '[Diprotodon] must be about'. In the water and at the edges of the water were objects that could be what we had been looking for, within 5 minutes I had located the weathered remains of 9 Diprotodon skeletons.[7]

Perhaps his equanimity derived partly from the expedition's mode of transport, by Land Rover and truck instead of camel, and from his experiences of similarly rugged conditions in field camps in the United States, Central and South America, but the rigours of a one-month journey were still considerable. The landscape provided compensation. He had rediscovered a site of great contemporary significance and palaeontological wealth, against the predictions of his colleagues at the South Australian Museum:

> It is almost fantastic to think that so many of these creatures bogged down here ... At present we have no idea how many ... It was a cheerful bunch that sauntered their way back to the Landrover and finally to camp ... most of the people around the museum really thought we wouldn't find the place. Others had looked for it and failed.[8]

Although in Australia principally to search for the elusive Tertiary remains, Stirton's joy in discovery, even of these younger fossils, is unmistakable: 'we hit bone and by evening had uncovered several vertebrae, ribs, two scapulae and most of the cranium. It is a pleasure to give a field number to a fossil.'[9] The landscape featured only in so far as it thwarted or facilitated science. It intruded as a mild inconvenience, a recalcitrant container unwilling to relinquish palaeontological knowledge: 'The mud is so sticky it is a fight to get it off the shovel … the water had seeped in again and we were standing in slush and water half way up our shoes … It is difficult to get close to the water because of the soft mud.' The fossils and the traces of past scientific endeavour were both features of the landscape – 'My favourite place is on the femur of a partly eroded Diprotodon exposed at the water's edge' – and measures of its stability – 'It was in a mound like this that I found one of the whiskey jars indicating [the] stability [of the mounds] for as long as 50 years.' Tedford was equally impressed, after Stirton surprised him, Woodard and Lawson with his discovery:

> The scene that lay before us was fantastic – there was Stirling and Zietz island camp – bottles, pottery, whiskey jugs, wood – were everywhere – and around this camp were the fossils that had made Callabonna famous in Australian VP – the gigantic Diprotodon – 9 skeletons lay before us in this saline waterhole only about 50 yards long – many more were to be seen in the flats north of the old campsite untouched in 60 years![10]

Recuperation and rehearsal in an 'heroic age' of geology

Despite their dominance of the first five *Memoirs of the Royal Society of South Australia*, in 1953 Stirton, Tedford and the University of Adelaide geologists seemed unaware of Stirling's and Zietz' monographs on the other Lake Callabonna fauna: the remains of *Phascolonus gigas*, a giant wombat, and of *Genyornis newtoni*.[11] The 1953 expedition found only Diprotodon fossils, though plenty of rabbit skeletons.

In 1958 Stirton wrote in some surprise that his attention had been directed to a large *Sthenurus* cranium 'in the old Stirling Zeitz [sic] collection from Lake Callabonna … Evidently the Stirling–Zeitz party found a site or sites where they discovered many things besides *Diprotodon*.'[12] The cranium of the giant short-faced browsing kangaroo had not been prepared or described and was 'in rather poor shape but I believe we can prepare it. Dr Brian Daily offered to let me ship it back to Berkeley for its preparation and description.' The preparation, preservation, identification, construction and publication of the Diprotodon, *Phascolonus*, and *Genyornis* remains evidently took so much time – in fact 20 years – and money that E.C. Stirling decided not to prepare or describe the sthenurine remains. In 1894 he had still hoped to do so, writing in *Nature* of the kangaroos that 'we have one small but very complete skeleton, and a large series of separate bones of several larger kinds, including a fairly compete skull'.[13]

Given the historical consciousness usually well-developed in vertebrate palaeontologists, it is surprising that no one at the museum or university directed Stirton's attention to the presence of other fauna at Callabonna until 1958. Perhaps they forgot, distracted by the peculiarly homogeneous assemblage at the Zietz–Stirling campsite, which produced great quantities of Diprotodon but no other megafaunal remains. The promise of other types of Pleistocene fossils inspired plans for a 1958 expedition to Callabonna. This was cancelled in favour of a promising Tertiary site at Lake Ngapakaldia in the Lake Eyre Basin, where on 25 July 1953, Stirton recorded with excitement the unprecedented discovery of a Tertiary diprotodontid fauna which was 'as abundant here if not more so than Diprotodon was at Lake Callabonna'.[14]

'Stumbling' across *The Dead Heart* in an Adelaide library, the 1958 expedition followed partly in Gregory's footsteps along the Cooper and Warburton rivers. Stirton's sense of history infuses the account in his journal: 'Upstream from where J.W. Gregory reported encountering a sandstone' the party came to 'Wara[l]amanka Waterhole, where Gregory's party weathered a sand storm followed by rain. Our weather is somewhat similar. Sand is blowing everywhere with showers from time to time.'[15] Stirton still lived in an heroic age of geology. He saw himself and his colleagues as explorer scientists, in the mould of men like Antarctic explorer Ernest Shackleton, Douglas Mawson and Gregory. Stirton and Tedford published popular articles on the central Australian and Callabonna expeditions that celebrated the Earth scientist as adventurer in the spirit of archaeologists Arthur Evans and Heinrich Schliemann in the Aegean, or even Indiana Jones. Gregory was himself the model, in a series of children's books, for war correspondent Osmar White's deranged Scottish adventurer–geologist, Dr A.A.A. McGurk, who set off on camel-back for the dread Deadibones Desert in search of the Super-roo of Mungalongaloo.[16]

These accounts, with their recursion, served a pedagogical and inspirational purpose as well. Stirton hoped his initial work in Australia would be 'responsible for rapid strides in the development of vertebrate palaeontology and continental stratigraphy in that fascinating land that still challenges the exploring scientist'. In 1955 he wrote to Herbert Hale, then the Director of the South Australian Museum, that he longed to be rid of his Chair in Vertebrate Palaeontology at Berkeley to concentrate on research and field work.[17] Stirton wrote of his desire 'to unveil the origin and succession of mammals ... one of Australia's greatest secrets'. He had 'dreamed for twenty years' of the Australian fossils, and conceived the 1953 trip while working on the Tertiary record of Colombia. His journal entry for 14 February 1953, the day he arrived in Sydney, expressed his relief and excitement: 'At last I had gotten to Australia in my quest for the ancestry of the monotremes and marsupials.'[18]

Pedagogy and scientific imperialism

Despite reservations about the universal applicability of George Basalla's model of the spread of Western science, the Berkeley connection to South Australian palaeontology highlights the importance of power relations between the palaeontological periphery and a new scientific metropolis. The ascendancy of North American universities and of the United States as an imperial power during the twentieth century saw a corresponding rise in scientific patronage in the 1940s and 1950s. The world became a site for knowledge to feed back into the US academy. However, Stirton saw an important part of his mission as fostering vertebrate palaeontology in Adelaide and the rest of Australia, as well as fact and fossil-gathering for the University of California. His self-assigned task in Australia was broader than a simple palaeontological treasure-hunt to fill in some of the missing pieces in the record of the descent of mammals. He sought to stimulate research in mammalian palaeontology in Australia and foster closer trans-Pacific, cross-institutional ties.[19]

With his pedagogical mission in mind, he gave public lectures, visited schools and spoke to the press. For example, in September 1954 he delivered a lecture to the Queensland Division of the Geological Society of Australia and the Field Naturalists' Club on 'The giant fossil marsupials of Queensland'. From his arrival in Adelaide in February 1953, he tolerated the constant media attention: 'Yesterday they took our pictures for the paper ... Today I was asked to give a recording for broadcasting. It will be good when all this is done with.' He was pleased to note on his return from Lake Callabonna in 1953 that 'Everyone asks me about Diprotodons. Apparently they are talking about it in most of the schools.' In a letter of May 1955 to Dick and Beth Tedford (the South Australian sweetheart Tedford met in Adelaide two years earlier, and married) in Oahu, where Tedford was stationed for his National Service, Stirton mentioned giving three seminars and three lectures on Australia during one week at the University of

Michigan. He also lectured on 'his Australian discoveries' in South Africa and the United Kingdom in 1958.

Stirton was a mentor to Australian geology graduates such as Geoff Woodard, who accompanied Stirton, Tedford and Lawson on the 1953 expeditions, and who subsequently received funding to complete his doctorate in vertebrate palaeontology at Berkeley. He brought American graduates to work on Australian material, such as Tedford and Les Marcus. He also found funding for South Australian Museum employees like Paul Lawson to study at the Museum of Vertebrate Palaeontology at Berkeley (UCMP).[20] He spent some time in 1957 in Washington DC at the National Museum (Smithsonian Institution) compiling a list of Holocene marsupials for comparison with the fossil specimens the UCMP teams brought back from Australia. He returned most of the type specimens obtained on the Lake Eyre expeditions to the South Australian Museum and the University of Adelaide, after examining them and taking casts at Berkeley, because he considered it important that local institutions retain such important fossils. In 1957 at a conference at the Smithsonian he recorded his desire to influence the Adelaide scientific establishment to 'do something serious for the future of Vertebrate Palaeontology' in South Australia, and spent some time lobbying his Australian colleagues.[21]

Stirton was not the only mammalian palaeontologist working on Tertiary remains in Australia. Australian palaeontologists like W.D.L. Ride were beginning to reassess Tertiary mammalian material during the 1950s and 1960s, and Ernest Lundelius, then a recent doctoral graduate from the University of Chicago, was independently looking at the Western Australian and Nullarbor records. But Stirton was the most influential and, despite his professed discomfort with interviews, an excellent science communicator and promoter. As well as facilitating the work of Americans on Australian material and *vice versa*, he collaborated closely with Australian workers in Australia, for instance Edmund Gill of the Museum of Victoria, who did early work on radiocarbon dating.

So Stirton was central to the emergence of a trans-Pacific palaeontological intellectual community, facilitated by that first cooperative venture between the South Australian Museum and the University of California Museum of Palaeontology's Fulbright-funded expedition to Callabonna in 1953. Stirton was committed to Australian vertebrate palaeontology in the long term. The serendipitous discovery of Tertiary koala remains at Lake Palankarinna by Geoff Woodard on 27 July 1953 was 'the first link in the chain of discoveries that should be followed through the next fifty years'.[22] Fifty-five years later, more than four decades after his early death of a heart attack, the 'chain' is indeed still being 'followed through', thanks in no small part to Stirton's legacy.

When Dick Tedford revisited Callabonna in 1970 after an interval of nearly two decades, the fate of Pleistocene megafauna was back on the intellectual agenda for American palaeontologists. The Smithsonian Institution, in the process of revising its fossil vertebrate collections, had funding for a Gallery of Pleistocene Giants from across the world. So, with the aid of a National Science Foundation Grant, it financed an expedition to Callabonna in collaboration with the South Australian Museum and the American Museum of Natural History (AMNH) in New York where Tedford was by then Curator of Vertebrate Palaeontology.[23] Although funding fell short of the eventual requirements for reconstruction and display, the expedition itself returned a bumper crop of extinct megafaunal remains and, as with the earlier expeditions to the lake, excited much media coverage in Adelaide.[24] The AMNH has a cast of one of two complete skeletons of giant wombats (*Phascolonus gigas*) recovered from the 1970 expedition to Callabonna displayed in its gallery of extinct mammals. There are no complete casts of this animal displayed in Australian institutions.

Figure 29: 'The Diprotodon amidst surroundings that represent some present characteristics of Central Australia.' *(Restoration of* Diprotodon australis, *C.H. Angas, 1907. Figure 2. Reprinted by permission from Macmillan Publishers Ltd: E.C. Stirling, 1907. Reconstruction of Diprotodon from the Callabonna Deposits, South Australia,* Nature, *76(1978): p. 544, copyright 1907. National Library of Australia S 505 NAT (1544216). Stirling commented: In the background, to the left, is part of the white expanse of one of those large salt-encrusted clay pans of which Lake Callabonna, where the bones were found, is an example. In the distance beyond the lake is shown one of the flat-topped hills that are very characteristic of the 'desert sandstone' region of the interior. The vegetation in the foreground is chiefly 'saltbush' (Atriplex spp.), some species of which, together with allied plants, having apparently formed the principal food of the Diprotodon, just as these now supply the chief sustenance of the introduced Herbivora, while here and there is a trailing plant of 'parakylia' (*Claytonia *spp.), so well known to travellers in the dry central regions for its moisture holding properties.)*

The South Australian Museum achieved Professor Owen's desideratum – a complete Diprotodon skeleton – in 1906. But there is an epilogue to the story of Lake Callabonna's Diprotodon graveyard. Not until nearly 70 years later, when 12-year-old Kerry Hine found the first undeformed Diprotodon skull at her father's clay quarry at Bacchus Marsh in Victoria (also the site of Permo-Carboniferous glacial remains like those at Hallett Cove) did the last, loose thread of the Diprotodon mystery knit into place (see Figure 29).[25]

Skeletons in the cabinet and the lost world of Lake Callabonna

What do stories about Diprotodon bones, dirt and the history of vertebrate palaeontology tell about the way the Earth sciences are made?

As at Hallett Cove, the multiplicity of geological heritages at Lake Callabonna begins in stories, read in each other, in the landscape and in its fossil productions: geological heritage as Earth history, painstakingly reconstructed by palaeoclimatologists, geologists and palaeontologists from its piles of dirt and lumps of bone; and as the history of vertebrate palaeontology embedded in its landforms, in its rescued skeletons and in its archive of field notes, anecdotes, borrowed ethnography and photographs. But while the fight for Hallett Cove was a test case for geological heritage in South Australia and has become emblematic of a perceived shift in environmental policy in the state during the 1960s and 1970s, public contest has not marred the frosted gypsum surface of Callabonna for a hundred years. Unambiguously an entrapment site and in any case long supposed to pre-date human occupation of the interior by up to a thousand centuries, it does not directly intrude on often heated contemporary debates about the causes of the extinction of the Australian megafauna.[1]

Its desiccated landscape illustrates the redemptive power of deep time and palaeoenvironmental reconstruction in a different way from Hallett Cove. In its evocation of dying Diprotodons in a dying lake, the standard palaeontological Just So story refreshes the trope of the watered inland in a Cenozoic land of plenty, reiterated in the Willandra Lakes story in Part Three. The interpretation of the material remains of Callabonna's Pleistocene past gives the modern landscape layers of meaning otherwise unintelligible and unimaginable.

Field work takes on different meanings at Callabonna and Hallett Cove. The slippage between laboratory and site are differently nuanced. Museum collections can stand in for and replace the physical site at Callabonna. The pragmatics of distance, collection and preparation ensure that it remains a lode to be mined, not a museum or teaching resource in its own right as Hallett Cove is. But the lake itself and its bones have a mystique lacking in the everyday utilitarianism of Hallett Cove's readily accessible outdoor laboratory. This mystique cannot be divorced from its scientific value and interpretation, suggesting that geology and the popular appeal of the deep past can be harnessed in the development of a landscape aesthetic appropriate to flat, dry north-eastern South Australia.

Geological heritage management is a poorly funded, barely legislated process of balancing the often competing needs of heritage preservation, industry and the extractive sciences: custodianship versus resource management. The GSA (SA) recognises Lake Callabonna and certain other of Stirton's Lake Eyre basin sites for their association with the development and articulation of matters of palaeontological and palaeobiogeographical concern in Australia. This heritage recognition in a way sits uneasily with Tedford's and Stirton's accounts of major excavations at Lakes Palankarinna and Ngapakaldia, and speculation about the future productivity of Callabonna. The scientists are nonchalant in print about ripping tons of sediments from the lakes' surfaces in the interest of the science. For instance, Stirton speculated in 1958 after finishing up for the season at Ngapakaldia that 'Much more could be found there by removing something like four feet of overburden, down to the mudstone. This could be done with a slip behind a landrover.'[2] Tedford's photographs in his contributions to *Kadimakara* and *Vertebrate Palaeontology of Australasia* demonstrate the scale of excavations during the 1960s.

This tension is more apparent in National Parks and in World Heritage areas, where scientists and parks staff occasionally find themselves in conflict. Mike Archer, a vertebrate palaeontologist and then the Director of the Australian Museum, was the subject of less than favourable press, having used gelignite to remove overburden at the Riversleigh World Heritage fossil site in Queensland in 2001.[3] Palaeontologists at the Naracoorte Caves, Riversleigh's South Australian partner, are sometimes frustrated by National Park officials who demand detailed reports before they are allowed to excavate. In turn, NPWS staff are anxious that landforms and fossils not be destroyed, removed or compromised without justification. Parks staff and site palaeontologists have worked together to produce a draft code of conduct for excavators.

Work at Callabonna has been erratic, due to its isolation from Adelaide and the nature of its sediments and climate, although this isolation provides a form of protection as well. The inaccessibility of sites such as Callabonna adds an extra dimension to site management in remote areas. Isolation does not always offer protection, as witnessed by the removal of valuable Precambrian fossils from the Ediacara site in the remote Flinders Ranges since the 1960s, and the theft of dinosaur trackways near Broome in Western Australia in 1996. The bones at Callabonna are very difficult to extract, the problem of their post-excavation preservation has not been solved and the region is nearly impassable during anything but the driest season. But its decoding has a genealogy reflecting international intellectual trends, patronage, parochialism and switching allegiances, personal and national. The desiccated landscape is beguiling, with this tantalising history, its palaeontological and palaeobotanical richness and the eerie vision of the behemoths of the marsupial world trekking across the surface and dying in droves.

Part Three

Lake Mungo, human antiquity and the watered inland: reading the scripture of the landscape

Most people think that Australian archaeology began in 1969 at Lake Mungo

David Horton, 1991

Beginnings

The Willandra Lakes region, since Jim Bowler's discovery of ancient human remains at Lake Mungo in the 1960s, has come to represent a fertile site for beginnings, a desiccated Garden of Eden, for archaeologists, heritage specialists and the Australian media. Unquestionably it has provided rich intellectual pickings. Lake Mungo is now a site of secular pilgrimage, a media darling and a favourite destination of the Qantas in flight magazine. And its most famous human relic personifies for Indigenous people a beginning and a link with their land. In the absence as yet of any evidence for conclusively older human occupation of Australia, it is tempting to indulge the powerful symbolism of the ancient burial Willandra Lakes Hominid (WLH) 3 (Mungo Man) as an Australian Adam and of WLH 1 (Mungo Lady or Mungo Woman) as Eve. Deucalion, son of Prometheus, and his wife Pyrrha, provide an even more resonant analogy, casting stones over their shoulders on the slopes of Mount Parnassus after the retreat of the floodwaters to renew the human race and with it usher in a

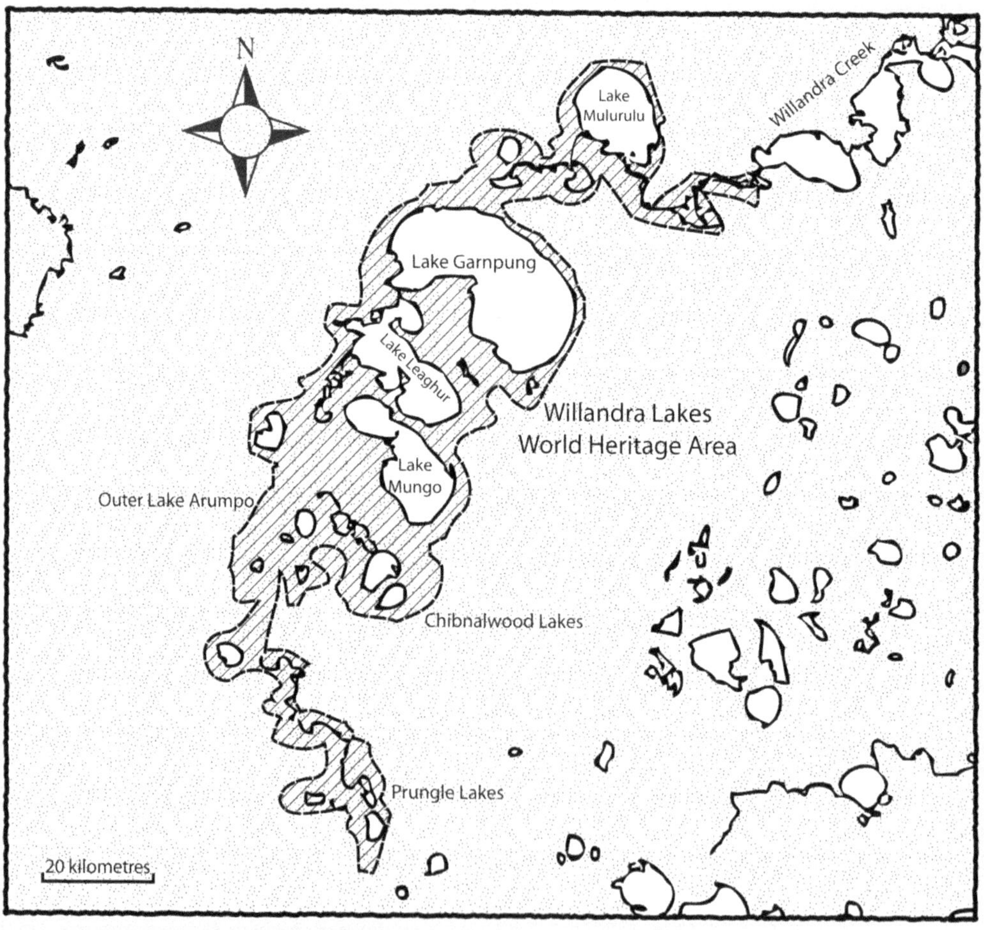

Figure 30: Willandra Lakes Region World Heritage Area. *(Image Brett Calcott)*

new golden age of archaeology in Australia. But in scientific terms, the region is as notable for its non-human relics and landforms, and what they represent in terms of climate change during the Pleistocene, as it is for its headline-grabbing, mythopoeic archaeological remains (Figure 30).

The device of beginnings foregrounds a number of recurrent and contrasting, often inimical, interests and themes that intersect in the stories of the Willandra Lakes. For the geomorphologist Jim Bowler, an interest in dry lake systems was the beginning of a spiritually and intellectually fruitful relationship with the region. This led to his felicitous discovery in 1968 of the Pleistocene human cremation, later known as Lake Mungo 1 or WLH 1, and its subsequent field identification as a modern human female the next year by John Mulvaney, Rhys Jones and Harry Allen, archaeologists from the Australian National University. Another beginning is represented by the successful nomination of the Willandra Lakes region for inclusion in the World Heritage List by the Australian Government in 1980. On 11 January 1992 Alan Thorne, a physical anthropologist, acting on behalf of Australian prehistorians, handed Mungo Lady back to Indigenous custodians at a ceremony at the Walls of China in Lake Mungo National Park, signalling a fourth beginning, one which perhaps includes improved relations between scientists and Indigenous people in the area.

That promise of a watered inland which led Bowler to the Willandra system has been rehearsed by scientists and explorers throughout the history of European and scientific exploration of the arid interior of Australia. When Charles Sturt and Thomas Mitchell travelled through (or in the case of Sturt, rowed past) this country between the Lachlan and the Darling Rivers, they saw in the arid landforms signs of past inundation. In common with the mediterranean obsessions of many of their fellow explorers, from Joseph Banks and Matthew Flinders to E.J. Eyre, they misread signs pointing to the existence of an inland sea, and hoped against the evidence for navigable rivers. But their interpretation of the geomorphology and palaeontology of the region as supporting models of former immersion resonates with modern palaeoclimatic and geological orthodoxies.

These models suggest alternating arid and humid regimes during the Pleistocene as ice sheets in the Northern Hemisphere waxed and waned, and point to the retreat of the sea during the Pliocene, which until perhaps a few million years ago lapped against the western plains of New South Wales.[1] In this sense, the palaeoclimatic baseline offered by the lake landforms continues a tradition of field interpretation which sees landscapes as encrypted text books, or scripture, from which deep histories and palaeoclimates can be deciphered, and the complex human history of the area may be interpreted.

The intellectual promise of the country is clear in Bowler's observation that 'It was amazing to find, in this arid land, abundant shells of freshwater mussels, bones of fish and beaches with shoreline gravels, dramatic evidence of times very different from those of today.' He described a region of contrast and boundaries: a dry place shaped by water; 'lake beds' awash in windblown dust and saltbush; freshwater mussels and fish bones embedded in desert sands. Its fragile landforms provide an index of climate change and landscape stability over tens of millennia. The poster-child for human antiquity in Australia was, until 30 years ago, a working farm. Deep time at Lake Mungo sits close to the surface, scoured out by wind and rain. Bones buried for thousands of years are exposed in minutes by wind or washed away by torrential downpours. World Heritage listing offers legislative protection, limited funding and the most effective tourism brand in the world to a site of great significance to Indigenous Australians, but it also, as the nomination document sought to demonstrate, links 'the origins of modern society in the Old World with one across Wallace's line in Australia': it is local and global in its intellectual implications. This region that has 'established the great antiquity and richness of Aboriginal culture' and 'caused a significant reassessment of Aboriginal prehistory' is thereby appropriated nationally and internationally as part of a project to preserve and catalogue sites of global heritage significance.

It is a place of demarcation. Bowler hints at one such line in the sand: 'My charter was geology and environment; this fell directly into the field of archaeology with all the sensitivities that involved.' The geomorphologist was still licking his wounds after heavy criticism from parts of the archaeological community, following his involvement in the Green Gully excavation of 1967. He elaborated during a conversation in May 1999, explaining that his first reports from the Willandra Lakes of dates of 30 000 years on mussel shells 'were greeted sceptically'. He was told that it was 'beyond belief' that the shells were human artefacts. Bowler, however, was 'convinced the shells had been carried into the dune by humans', but he 'was not an archaeologist', and it was suggested he go back and 'find something that was more credible'.

The discovery then of those bones 'sticking out of a lump of calcrete' which 'were the bones of a large mammal and they were burned and they were smashed' provided conclusive evidence of human presence for Bowler. But, when working in the Keilor terraces west of Melbourne with geologists Tom Darragh and Dermot Casey 'only a couple of years before', the three had removed skeletal material from Green Gully 'and we were chastised' for removing these materials 'without "proper" authentication'. Having 'had my fingers burnt', Jim was unwilling to bring unauthenticated remains back from the lake without their being 'seen by archaeologists in situ'. So until excavated and identified by archaeologists, Mungo Lady's bones 'lay awaiting expert appraisal'.[2]

The needs of Aboriginal people and of scientists also come into conflict, as in 1988 when the Western Regional Land Council placed an embargo on archaeological excavation at Lake Mungo. A path to partial reconciliation of such apparently incompatible worldviews is illuminated by the repatriation of Mungo Lady in 1992. Those who believe ancestors belong in the ground and those who believe that knowledge of the dead belongs to all mankind, or at least to those who are prepared to excavate and publish, came together on a day in January that year, when a physical anthropologist handed the bundle of bones to local elders from Balranald, who now store her in a secret keeping place, from which she may be retrieved at some unspecified future time.

The acquisition of Mungo Station from Albert and Venda Barnes for the creation of the Lake Mungo National Park in the 1970s represents another line in the sand, between archaeologists, pastoralists and Aboriginal people. Although the Barneses regarded the park as their legacy to the nation, some pastoralists fear that the discovery of human remains on their leasehold will lead to similar reclamation. Some of these concerns were allayed when the boundary of the World Heritage Area was redrawn in 1996 from more than 600 000 hectares to just under 240 000 hectares, excluding much pastoral land, and better reflecting the location of identified World Heritage values.

The Willandra Lakes Region World Heritage Area can be read as a place of boundaries or interfaces. The now dry lakes lie at an ecological interface, between the mallee-covered linear dunefields of the western Riverine Plain, formed principally by aeolian (wind) processes, and the fluvial (river) landscapes of the eastern Riverine Plain. Their surface expression has been carved as much by wind as by rain and rivers, their sediments influenced by subsurface hydrological regimes as much as meteoric water. Essentially dry for upward of the last 12 000 years, sited on the Willandra Billabong Creek, a long-ineffectual outlier of the Lachlan River, these 'lakes' represent a geomorphic–semantic contradiction.

The area is at a number of cultural and intellectual interfaces. Since Bowler, then a doctoral student in geomorphology at the Australian National University, decided in the late 1960s to investigate the lakes' potential as palaeoclimatic indicators for the Quaternary period, they have come to represent windows onto a more humid, climatically variable past, in dramatic counterpoint to the apparently timeless and changeless semi-arid modern landscape. The lake and lake shore sediments preserve evidence of cyclic climate changes between wet and dry regimes over the last several hundred thousand years as ice sheets waxed and waned elsewhere, vividly demonstrating the contradiction of arid, iceless 'glaciation' in low latitude, low altitude Pleistocene Australia. The sediments, flora, fossil fauna and record of Aboriginal occupation recall patterns of fire and flood, drought and river-flow for a hundred or a hundred thousand years. On a shorter time scale, the lakes display the impact of agriculture, pastoralism, introduced pests, and catchment and groundwater changes relating to water management and irrigation since European settlement.

In July 1968, due to a combination of serendipity, a sharp eye and his eagerness to support a conviction of Pleistocene human occupation of the lake shore, Jim Bowler climbed off his motorbike to investigate a bundle of cracked, charred, carbonate-encrusted bones in a disintegrating calcrete block. The following autumn an excited group of Earth scientists, including archaeologists Allen, Jones and Mulvaney, identified the bones as a human cremation and packed them into Mulvaney's suitcase (which is now in the collection of the National Museum of Australia) to be radiocarbon-dated at the Australian National University; Mungo Lady made her national debut; and the time of Aboriginal occupation of the continent deepened once more, to 25–32 000 years before the present. In February 1974, in another felicitous conjunction of pluvial denudation and Bowler's motorbike, the exhumation of Mungo Man – the oldest recognised ritual burial in the world – pushed Aboriginal occupation of the south-east back to the limit of carbon dating and confirmed Lake Mungo's status as a canonical archaeological location.

In 1979 Mungo Station was dedicated as a National Park, which incorporated the Zanci pastoral lease in 1984. The lakes' investiture as a World Heritage area occurred in 1981, and the boundaries were redistributed in the mid 1990s. In 1992 the Australian National University returned Mungo Lady for safekeeping to Aboriginal people who claim descent from her contemporaries. Parts of adjoining leasehold blocks have been added to the park subsequently. More mild controversy accompanied the re-dating of Mungo Man in 1998 and 1999 to 45–80 000 years. Since 2003, more than 450 human footprints dating to 20 000 years ago have been found on ancient surfaces near the lake shore. In 2009 the three traditional tribal groups (3TTGs), the Paakantyi, Mutthi Mutthi and Ngyiampaa people, and the New South Wales government agreed to joint management of the park. The answers to the question of human origins and longevity in Australia have implications for archaeological and geological inquiry in Australia and the rest of the world. It seems that age matters in western New South Wales.[3]

With particular regard to the establishment of Lake Mungo National Park, in 2002 Jim Bowler acknowledged the Barnes families, pastoralists on the former Mungo and Joulni Stations who 'endured the climatic and economic vicissitudes of life in the Western Division of NSW, lives frequently compounded by the complexities that followed on the heels of scientific discoveries and subsequent heritage complications'. He insisted that 'the nation in general' owes 'a profound debt' to the Barneses who 'forfeited their heritage, Mungo Station, to permit the investment of that region's scientific treasures in the national interest'.[4] The deep past and geological heritage thus redeem human and administrative failure in the recent past.

The Willandra Lakes and Lake Mungo in particular have been integrated into a corpus of nationally significant landscapes for their natural and cultural values. Fossil landscapes and bones in the lakes are vehicles for representation. The signs of the past, including human pasts, are central to the 'nationing' of Lake Mungo and its associated material remains. They become repositories of ideas about antiquity and also vessels for the articulation and development of national identity.

Chapter 8

ARID GLACIATION AND THE LAND OF LAKES

The Willandra Lakes Region World Heritage Area demonstrates at least 40 000 years of human exploitation. Thirty-two pastoral leases controlled by the Western Lands Commission lie partly or wholly within the region. It has been avoided, maligned, farmed, ignored, named, photographed, gazetted, painted and studied. Stories have been told about it. Its eponymous lake basins are dry, described as fossils. It is celebrated for landscapes that are the product of erosion. This has implications for land management, geological conservation and interpretation, as well as semiotics: What is a 'lake' in such a context?

This land of lakes is dusty country on the semi-arid fringe separating the mallee dunefields and the eastern Riverine Plains of New South Wales. Its lakes are mostly ovoid clay and sandy plains, with sparse salsolaceous vegetation. Sheep, kangaroos, emus and feral goats traverse their floors, treading down shells and the bones of fish. Its landforms helped geomorphologists and climatologists articulate a new understanding of the nature of the Pleistocene ice age in low latitude, low altitude mainland Australia, a nonglaciated 'glacial' period that resulted in counter-intuitive, dusty, cold aridity, as the deserts expanded, the lakes dried and the oceans shrank. In such an Australia, Sydney Harbour became a delta, Bass Strait a swampy plain joining a glaciated Tasmania to the mainland, and the Gulf of Carpentaria was a giant marshy corridor to New Guinea, raising questions about boundaries: national, climatic, temporal, ecological and physiographic.

Fossil landscapes

The allure of the ancient human presence in the Willandra Lakes region often blinds us to the wonderful record of Cenozoic climate change across south-eastern Australia represented by its landforms. The World Heritage Committee recognised the lakes as a site of natural as well as cultural patrimony.

The landscapes encountered by Paakantyi, Mutthi Mutthi, Ngyiampaa, returned soldier, pastoralist, explorer and scientist were carved, and continue to be shaped, by climate fluctuations over hundreds of millennia. The antecedents of the 3TTGs lived through some of these changes and left material for those who came after to question and try to understand. Traditions from elsewhere in the Murray–Darling basin and in the arid zone have been offered as analogies for the ways that the Mungo people might have understood the land. The experiences and observations of pastoralists suggest that often they were too concerned with survival to notice the lake beneath their feet, but there were exceptions. The reports of nineteenth-century surveyors and explorers, and the observations of artist Russell Drysdale and jounalist Keith Newman in 1944, show that the later discoveries and speculations of geomorphologists, archaeologists and palaeoclimatologists are not simply a product of late twentieth-century scientific savvy. Geomorphological and archaeological orthodoxy is often a rehearsal or refinement or secularisation of non-secular scientific ways of understanding country, a recurrent theme and key argument of this work.

The modern Murray–Darling Basin is a flat-lying depression that hugs the eastern margin of the Great Divide, spans central and western New South Wales, through Victoria and eastern South Australia to the sea. It encompasses over a million square kilometres, which is more than one-seventh of the continent. Apart from minor leakage and discharge via the Murray River to the sea, the basin is a closed groundwater system. Groundwater at the margins of the basin is less saline than that at the centre. The basin demonstrates a significant climatic gradient. Precipitation dominates to the east. Evaporative processes are more significant in the west. Prevailing winds are westerly. Surface water flows from the east. Seasonal contrast produces hot dry summers and cooler winters with rare frosts.[1] But conditions in the region were very different in Panel 6, during the mid to late Miocene and early Pliocene, when Hallett Cove may have been covered by a warm shallow sea and flamingoes and freshwater dolphins still lived in the waterways of the Lake Eyre basin.

Panel 6

Sedimentary sequences across the basin suggest eustatic sea-level changes during the last 20 million years. Non-marine sediments predominate in the north and the east, while sedimentary successions in the south and west extensively overlie marine deposits. Marine deposition persisted in the south of the basin until the upper Miocene–Pliocene, when the Tertiary sea retreated to the south-west, depositing the water-saturated Loxton–Parilla sandsheet. This sandsheet underlies much of the Mallee region and has significantly influenced ensuing Quaternary deposition. The sea covered the future site of Lake Mungo, but probably did not reach as far north as Lake Mulurulu, the northernmost lake in the chain. A series of sub-parallel, relict strand-lines of the Loxton–Parilla sands indicating former shorelines are preserved as north–south oriented ridges in the south and west of the basin and provided a structural constraint on drainage and lake distribution.[2]

The characteristic moon-shaped leeward dunes – lunettes – on the west of the Willandra Lakes probably formed on these ridges, except the Mulurulu lunette, whose core is a gypcrete ridge, possibly formed in an ancient hypersaline lagoon. Miocene-age events are separated from Pleistocene successions by a depositional hiatus that Jim Bowler has correlated with a deep-weathering period, as Australia drifted further north and dried. This might have closed at about 2.5 million years before present, at the end of the Pliocene.[3]

Early tectonic damming of the proto-Murray River led to the widespread early Pleistocene development of lake systems across much of south-west New South Wales and northern Victoria. The Blanchtown Clay and Bungunnia Limestone lake deposits provide one of few absolute temperature–time correlated controls for the region. These grade laterally into the eastern, unconsolidated silt and clay of the Shepparton Formation, which produced the low-relief Riverine Plain of today. Saline lacustrine–aeolian sedimentary assemblages in the west of the basin characterise the modern semi-arid Mallee region.[4]

Panel 7

In contrast to modern eastern Australian climates, most of the Cenozoic era was marked by relatively stable, humid climates, little or no Antarctic ice cap and higher sea levels. Although Australia was further south than today, the climate was conspicuously warmer at equivalent latitudes. During the Neogene, major changes in global circulation overwrote the relatively slow northward movement of the Australian plate. As ocean circulation changed with the opening of Drake's Passage and the circumpolar Southern Ocean, subtropical high-pressure belts intensified and migrated towards the equator, overtaking Australia from the south. During the Quaternary period, fluctuating humid and arid conditions were accompanied by profound hydrologic changes. These changes have left many of the distinctive landforms of the Murray basin which can be used as an archive of change for palaeoclimatologists.[5]

In 'Aridity in Australia: Age, origins and expression in aeolian landforms and sediments', an influential article published in *Earth Science Reviews* in 1976, Jim Bowler observed that in order to speculate usefully on past aridity, it is necessary to identify possible constraints on a swing to drier conditions. The major controlling parameters he considered were increased evaporation and decreased precipitation. Increases in evaporation are usually associated with increased temperatures. This appears at odds with observed Pleistocene glacial trends in Australia. However, increased wind strength can also effect an increase in evaporation, and an altered aeolian regime would not be unexpected under glacial conditions. Therefore it is necessary to isolate changes in precipitation, temperature and wind regimes.

Bowler distinguished four entwined lines of primary evidence:

- *biological*: vegetation, human and faunal assemblages suggest increased aridity;
- *hydrological*: fluctuations of lake levels in the late Neogene and Quaternary periods are inextricably linked to climate. This is reflected in lake and river stratigraphy. Abandoned or reduced rivers reflect hydrologic changes, as observed in the region during the previous century by Charles Sturt, Thomas Mitchell, and recorded later in South Australia by Ralph Tate and others. However, fluvial data are more difficult to constrain and therefore are subject to more ambivalent interpretations than the evidence from lakes.
- *aeolian sediments and landforms* such as dunes and dust mantles: their increased quantity and mobility during the terminal Pleistocene epoch points to greater aridity accompanying glacial conditions.
- *soil data and stratigraphy*: soil formation and preservation alternates with fluvial and aeolian deposition, and the deep regolith profile of the Mallee–Riverine Plain preserves buried soil successions. Such preservation of pedogenesis is therefore a surrogate for altered wind–water regimes.

Data obtained from stratigraphic and geomorphological studies in the Murray basin suggest that large lakes and rivers were active prior to 50 000 years ago, before the last glacial maximum (the period of maximal ice sheet advance in the Northern Hemisphere, which corresponds to the driest conditions in Australia). As water levels started to fall, lake drying led to deflation of clay from lake floors and the consequent development of arcuate dunes (lunettes), which then evolved red soils when conditions stabilised. Increased precipitation versus evaporation between about 50 000 and 25 000 years ago induced water table rises and a subsequent freshwater, lake-full period before about 35 000 years ago. Stratigraphic evidence for this includes sandy beaches and foredunes on the eastern shores of the lakes, and the striking evidence for human occupation, in the form of middens, hearths, tools, a silcrete quarry at Lake Mungo, and human burials with sophisticated funerary rites. Streams exhibited larger channels, meander wavelengths and widths, with an estimated three to six times the volume of present streams. From about 25 000 years ago, the development of sandy point bars and source-bordering sand dunes along river channels indicates remobilised sediments and attests to the sparseness of vegetation, which would have otherwise stabilised the landscape to some extent.[6]

The last glacial maximum, characterised in inland Australia by aridity and low lake levels, began about 25 000 years ago. Small glaciers formed in parts of Tasmania and possibly the top of the Kosciuszko Plateau, which experienced periglacial conditions. Otherwise, Australia remained free of permanent ice. Less precipitation in the form of snow in the high country reduced spring run-off to rivers and lakes in the eastern inland, exacerbating the effects of the drying. Periods of peak aridity and increased salinity fell between 21–18 000 years ago, exacerbated by strong westerly winds and sparse vegetation, but showing local variation. These conditions can be deduced from the contemporaneous occurrence of variably gypseous clay lunettes around depressions or on the eastern sides of lakes. Subsequent grey soil pedogenesis, deposition of parna on the Riverine Plain and remobilisation of calcareous mallee dunes suggest strong wind regimes and a cessation of extensive fluvial deposition. People of the region might have dissolved into smaller groups, or been driven up-basin, first to Lake Mulurulu, which contained water longer than the southern lakes, and then to the rivers and coasts.[7]

Panel 8

Following the initiation and development of modern rivers, the lakes remained saline. Evidence from the Willandra Lakes and Lakes Tyrrell and Keilambete in western Victoria suggests water levels began to rise again from about 14–12 000 years ago, and reached their maxima between 8–6000 years ago, before a slight decline throughout the rest of the Holocene, but regional variation can be distinguished. People probably returned seasonally to the lakes region, exploiting the grain crops, birds and mammals of the semi-arid zone, rather than the previously abundant freshwater harvests. Eucalypts stabilised riverbanks, flood plains, dunes and lunettes. Lake basins generally remain dry except where the water table has been raised by irrigation.[8]

'A site un-named is a site unknown': exploration and pastoralism in the Willandra Lakes region

The story of European exploration and pastoralism in the Willandra Lakes region belies Jim Bowler's claim that 'a site un-named is a site unknown'. While the Indigenous Barindji people probably had names for specific landforms and for the area as a whole, to the first European settlers, the Lake Mungo basin was simply part of Mungo Station, established in 1864. The fact that some of its features did not enter Anglo-Australian scientific consciousness until more than a century later, when Bowler named the lakes after the pastoral leases into which they were incorporated, did not prevent them from becoming objects of contemplation and speculation earlier.

Thomas Mitchell never went inland from the rivers as far as the Willandra Lakes, but had he done so, he would almost certainly have recognised them as dry lake basins, based on his experiences nearer to the Murray River and in north-western Victoria. In June 1836, his party 'passed along a low sandy ridge, every way resembling a beach, but covered with pines and scrub. A bare grassy hill extended southward from each end of it.' The 'interesting hollow' to the west contained some water, and was 'evidently the bed of a lake, nearly dry'. Mitchell 'was struck with the analogy, and in these ridges being always on the eastern shore of hollows or lakes' while 'the western was irregularly indented, and was in some parts so abrupt as to have the character of cliffs'. He observed other basins, frequently waterless or saline, with similar features: 'nearly all circular or oval' with 'a very regularly curved ridge', now named lunettes. Further, he noted 'native ash-hills', ovens or middens on the sand ridge, which at least proved 'that this lake had once contained' mussels, suggesting fresh water in the past. He had observed fish and mussels 'cooked in such ovens as these' elsewhere on the Murray. Although fascinated with, and conscious of the utility of, local Aboriginal names for places and landmarks, he rarely asked the names of these basins, nor did he name most of them himself, though he recognised their lacustrine nature.[9] Charles Sturt, closely followed by Mitchell, was the first Englishman systematically to explore the central and southern reaches of the Murray–Darling basin. Both men saw in the landscape signs of wetness, signifying to the latter greater fertility in the past and to the former, an extant inland sea. Such ideas resonate with modern secular geological models of the vanished world of the Pleistocene and the interpretation of Lake Callabonna during the late nineteenth century.

Although it is commonly asserted that Sturt introduced cattle to the region in 1828, it is more precise to suggest that he *lost* cattle in the region in 1828. Thomas Mitchell's experience of Darling River people who, to his mind, must 'never' have seen cattle, but who evinced no surprise at their appearance, indicates that in any case they could have been present for some seven years since Sturt. The first European to survey near the lakes region was probably Granville Stapylton, Mitchell's ill-fated and erratic second-in-command on the Australia Felix expedition, who examined country west of the Lachlan in May of 1836 before he was killed. In 1860 Burke and Wills rested at nearby Arumpo Station on their way to death and immortality.

The history of the region is a history of the inextricability of deep from shallow time in modern Australia. The wool shed at Mungo Station was built by Chinese labourers in 1869. The Walls of China, a distinctively eroded part of the lunette on the north-east shore of Lake Mungo, may have been named for these labourers, but this is not documented. By 1877 Gol Gol Station covered more than 200 000 hectares. Oral and written reports by local graziers and government land inspectors suggest that, like Mitchell, many pastoralists and surveyors recognised the large saltbush and grass plains within the region as dry lakes, formerly filled by the Willandra Creek. In 1998, archaeologists Harvey Johnston and Peter Clark identified land inspection reports made on Gol Gol Station from 1875 to 1900 that refer to areas of open plain and adjacent dune ridges as former lake beds, and a series of 'hills' – the 'Gal Gal Hills' – which may correspond to the Walls of China. Thomas Mitchell's 1838 text does not mention, but his map depicts, the 'Golgol Range' in south-western New South Wales, which lie approximately in line with the position of the lakes, and may represent these Gal Gal Hills, although he oriented them incorrectly for the Willandra lunettes (see Figure 31).

Ongoing large-scale erosion and sweeping dust storms in the twentieth century resulted from the combined efforts of rabbit plagues, a succession of droughts from the 1890s onward and overstocking by desperate or inexperienced sheep farmers. This rapid denudation, exposing Pleistocene stratigraphy, perhaps aided the eventual discovery of the human burials. In 1921 the highly degraded Gol Gol Station was subdivided into 16 000 hectare holdings for soldier settlers. These subdivisions included Zanci, Mungo, Leaghur and Joulni Stations. The Cameron brothers, Ewan and Nagus, ran Mungo as a pastoral station from 1922 until 1934 when Alex and Albert Barnes, brothers who operated Joulni Station, acquired it. Although appreciative of the revelation of an ocean beneath their feet when Bowler explained his theory to them in the 1960s, the Barneses expressed no previous awareness that they lived on the floor of a fossil lake. However, their contemporary, Cliff Smith of nearby Chibnalwood Station, greeted Bowler's enquiries about unlikely shell middens and fish bones with enthusiastic recognition, having already, in Bowler's words, 'read the scripture of the landscape'.[10]

In a story paralleling the experience of the Ragless brothers half a century earlier in South Australia, in the 1940s farmers again fell victim to those unpredictable and seemingly capricious climate cycles. Rain stopped falling. Australia's eastern seaboard dried up and wind blew the topsoil of the interior to the coast and across the Tasman to New Zealand in great clouds of dust, famously staining the snow of the South Island red. The drought was considered the worst in Australia's history, until similar claims for the El Niño event of 2002 and 2003. In 1944 the artist Russell Drysdale and a journalist, Keith Newman, were commissioned by the *Sydney Morning Herald* to record the effects of this drought in western New South Wales, following the lifting of wartime censorship. The two, accompanied by a driver and J.M. Leonard, a photographer, drove several thousand kilometres through the state's interior in November and December. Newman's articles appeared in the paper on 16, 18 and 19 December accompanied by Drysdale's line illustrations and Leonard's photographs. A further article by Newman, with photos by Leonard, appeared on 20 December.

Figure 31: 'Golgol Range'. *(Excerpt from map of south-eastern Australia.* In *Mitchell, 1838)*

Artistically and personally, the experience was tremendously important for Drysdale. His biographer Lou Klepac wrote of Drysdale's experience of the Walls of China that 'Sitting beside the fire on a brilliant moonlit night, he found it one of the most extraordinary places in the world.' His drawings of eroded country, dead trees, stoic families, ruined houses and windmills, and bones, 'were seen at breakfast tables in households throughout the country'. Perhaps even more importantly, he produced a series of paintings from the sketches and photographs, including 'Walls of China, Gol Gol'. He brought the devastated interior home to the cities, and in the process he had a transformative effect on the places and people. 'Commonly thought inappropriate and ugly', he depicted them as tragic, their ruined fortitude providing a sort of horrible majesty – 'scenes of some grandeur' in 'a most moving poem of destruction'.[11] Historian Tim Bonyhady has pointed out that the celebrated if idiosyncratic archaeological foundation story of the lakes – geomorphologist on motorcycle finds human burial while looking for something else, archaeologists identify and collect it in a suitcase, saving it from washing away in the next unexpected downpour – somewhat overstates the role of serendipity. Both Newman and Drysdale conceived and represented the landscape as ancient and Aboriginal a quarter of a century before Bowler's motorbike confirmed it for the academy. With an echo of Mitchell's 'native ash hills', Newman wrote of the Walls of China that:

> The local people say that the locality is rich in Aboriginal relics. Burial grounds and weapons, and perhaps, bone fragments of the things the ancient people ate, are there to be found – and may be they could say whether one tribe flourished

> here for aeons of time, or whether spells of aridity drove them away until better seasons restored the water and the game.[12]

As Bonyhady suggested, anyone who read Newman's article could reasonably surmise that the Walls of China was as good a place as any in western New South Wales to find Aboriginal bones and artefacts. Furthermore, 'Any of Newman's readers would also have understood that to delay such digging was hazardous, since bones there one day might be gone the next.'[13] But Bowler came to the lakes not in search of human antiquity, but to understand climate cycles of the last hundred millennia, guided there by the reflections about saltpans by his mentor and supervisor, Joe Jennings, who noticed the Willandra Lakes during a plane ride over western New South Wales. The same conditions which make the preservation and discovery of human remains probable facilitated Bowler's geomorphological work. For Jim, at least initially, the evidence of Pleistocene humanity in the landscape was simply another indicator of ecological change.

Newman also understood that the 'flat grey depression' beneath the Walls was a dry lake, and gave a rudimentary account of lunette formation: 'this sandheap, unlike others, could not have been caused entirely by wind erosion. Some say the sand was on the lakebed and merely blew into this great heap when the water went. If that is so, the river that fed this lake must have been a mighty fellow ... He must have chewed up a province to spit out this gritty residue.'[14] Clearly, current geomorphological orthodoxy regarding the lakes and the formation of their leeward dunes is not new. Though Newman's account could do with some refining, it is not entirely incompatible with the mid-century explanations of geologists. On 20 December 1944, his understanding of the building of the Walls of China lunette was a little more scientifically shaky. He suggested the dune was an artefact of post-settlement erosion rather than evidence of the larger climatic cycles of which he was well aware: 'as this [vegetation] was thinned by axemen, drought, and stock, the winds concentrated on the remainder'.[15] He was not alone in neglecting the signs of older cycles of erosion. The Lake Mungo National Park plan of management made the same assumption in 2006:

> The accelerated erosion of the Mungo lunette which was largely brought about by European pastoral activity and introduced species such as rabbits has sculpted the lunette to its present shape.

Journalist Sandy Guy, writing about Lake Mungo for both the *Age* and for the Qantas in flight magazine in 2003, also misunderstood the nature of lunettes, assuming that the Walls themselves are a series of lunettes, eroded out of sand dunes surrounding the lake, rather than built from sand and clay blowing from the lake floor: 'Centuries of wind and rain have eroded the ancient sand dunes surrounding the lake, producing hardened, crescent-shaped walls of sand and clay known as lunettes.'[16]

In 1997, recalling Newman's vision of erosion at the Walls of China lunette as an artefact of livestock, and in a twist on the nineteenth-century notion of cultivation rejuvenating marginal country, science writer Mary White described Lake Mungo as a degraded landscape, ruined by pastoralism.[17] She, too, attributed the mobility of the landscape to overstocking, and asserted that this mobility threatens its aesthetic and scientific value. However, geologists read cycles of climate change and discover archaeological material in this badlands erosion, which itself is a source of aesthetic appreciation. In a landscape with as little natural exposure as western New South Wales, geological research relies to some extent on human intervention. As the authors of the management plan observed, 'This erosion uncovered the treasure trove of materials which made "the Walls of China" so well known and precipitated its World Heritage status as well as creating a scenic landform.' More recent scientific accounts of the Willandra Lakes landscapes suggest that even these erosional patterns – which facilitated geological understanding – are evidence of older climatic cycles at work. Sheep, goats and rabbits have undermined the stability of the lunettes, but in a repeat of patterns found deeper in the stratigraphic record. The modern Mungo lunette was not superimposed over a smooth underlying topography (see Figure 39, page 155).[18] These are stories the world tells about itself: beneath the decomposing land are the relics of its forerunners. Newman, in contrast to White, read a poignant message of hope in these signs of a former world: 'Does this mean that some time after the lake had died, another great cycle of rains packed down the sand and germinated plants, and that Nature with infinite patience had time to enrich the soil again before the droughts returned?'[19]

Landscape, antiquity and nationalism

Lunettes are part of the scripture of the land, both a turn of phrase that demonstrates the transformative function of language on landscape, and a means by which change over time can be 'read'. During the 1880s, surveyors' reports in western New South Wales noted the presence of collections of bivalve shells and fish bones around the rims and on the surfaces of broad dry ovoid saltbush plains.[20] As early as 1838 Mitchell identified these plains and their leeward arcuate dunes as dry lakes, by analogy with existing salt lakes:

> The hills on the margins of the Australian salt lakes are always on the north-east side, or opposite to that of the prevailing south-west winds. The formation of these hills is probably due to the action of the wind ... while in some cases, where grassy flats had once been lakes, crescent-shaped green mounds were still apparent on the north-eastern sides of each.[21]

The scientific salvation of the Willandra Lakes region lies in the more recent past. It was not until the 1940s that Melbourne geologist E.S. Hills described the formation of these 'crescent-shaped green mounds', or lunettes, as being driven by cycles of aridity and humidity and prevailing winds during the last ice age.[22] Lunettes are transverse, stable dunes which form upwind of lakes (see Figures 32, 33 and 34). They are ubiquitous in semi-arid southern Australia on the north-eastern shores of lake basins. They have also been reported in arid inland Australia where they may be critical elements in the genesis of longitudinal sand dunes. A number of models have been generated to explain their existence and development. The discovery of the diagnostic importance of ephemeral and dry lakes has allowed their reconfiguration as objects of natural beauty. Jim Bowler, writing about Australian salt lakes in 1981, suggested that they are 'as rich in scientific potential as in the beauty they often display ... even the dry and apparently desolate salt encrusted 'lakes' (salinas) of inland Australia possess their own harsh but enticing appeal'.

Among the lunettes of the Willandra Lakes, Bowler and colleagues from the Australian National University differentiated three main dune-building stages (Figure 33). A quartz sand foredune phase corresponds to lake-full, freshwater regimes and relatively humid conditions. A change to more arid, saline conditions and lower lake levels, when the watertable intersects with the surface, is represented by pelletal clay and gypsum, or copi, dune phases. Finally, pedogenesis (soil formation) occurs. Such soil disconformities provide evidence of an abeyance in active dune formation and deposition. They often separate major dune-building units. This well-documented sequence enables effective palaeo-climatic and hydrologic reconstructions.

Lunettes are bow-shaped in plan with fixed arms that point upwind, clinging to the shoreline. The crest of a lunette is closest to the lake edge, as the windward side of the dune is the

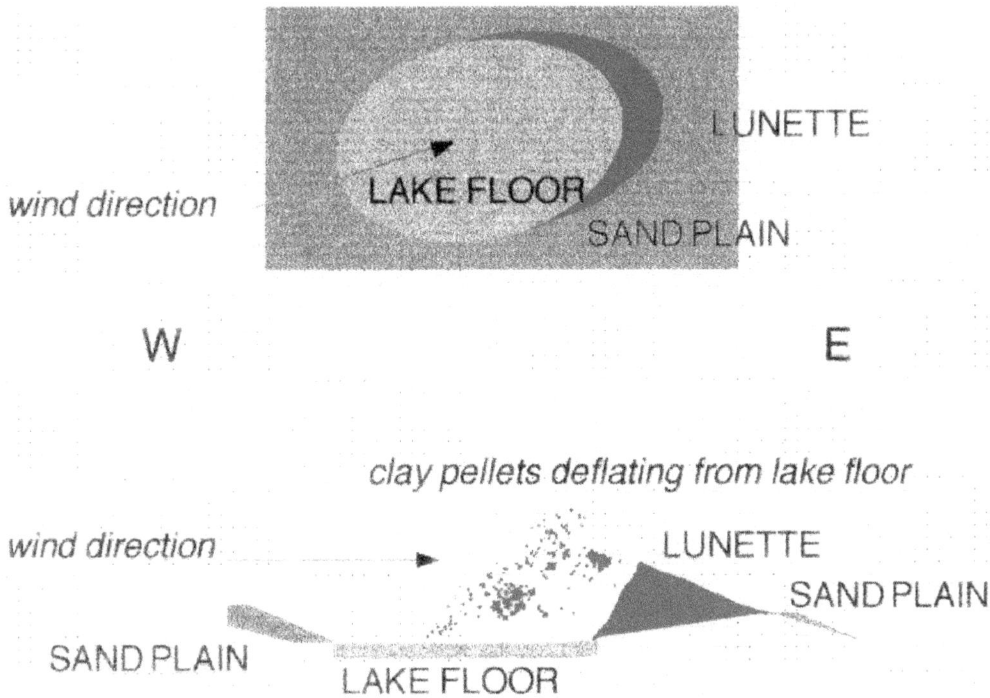

Figure 32: Prevailing wind direction shaping the formation of a typical lunette.

steeper of the two (see Figure 32). The size of a lunette is directly proportional to the size of the lake that it borders. Australian lunettes vary in topographic expression and composition, from silt and clay to gypsum to sand to a combination of these, and regularly exhibit cross-bedding or gravel. The Lake Garnpung lunette, north of Lake Mungo, is the biggest clay lunette in the world (see Figure 34). Dry lake lunettes are generally inactive, but the lunette at saline Lake Tyrrell in western Victoria is active, and may be a useful model for the formation of clay-phase dunes in the Willandra Lakes.

The Western Division of New South Wales has always been marginal pastoral country. Patterns of severe drought characterise the region. But perhaps the acceptance of limitation is contingent not so much on the pronouncements of climatologists and agricultural scientists, as on those constraints being brought home – literally – in the dust storms that hit the eastern sea-board sporadically when the western interior has been razed by drought, as in February 1983 and December 2002, and in gruesome photographs of dying and dead stock in the popular media. In the drought of the mid-1940s, this realisation of the limits of country was also dawning. Keith Newman observed that 'Nature's purpose was to hold the desert in check, not to provide food for imported animals recklessly multiplied.'[23] The *Sydney Morning Herald* commissioned the distinguished geographer Griffith Taylor, by then based in Canada, but formerly at the University of Sydney, to write a series of articles in 1944. His first was about Australian deserts, illustrated with a drawing by Russell Drysdale, of erosion south-east of Broken Hill. Taylor noted with irony, and not without bitterness, that his books had previously been banned in Western Australian schools for his views: 'There was universal objection to using the term "desert" for any part of the Australian interior. I imagine that Australians now realise that a desert by any other name is just as dry.'[24]

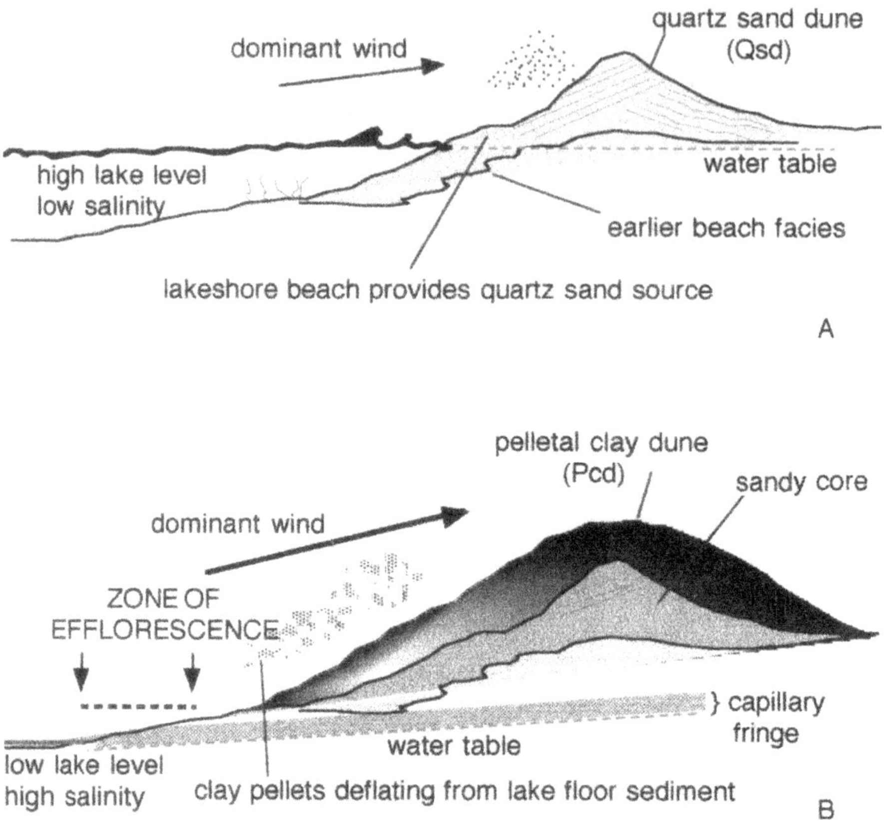

Figure 33: Formation of a lunette.

Figure 34: Lake Garnpung in the Willandra Lakes has the largest known clay lunette in the world. *(Photograph Michael Main)*

This new-found ecological sensitivity that Taylor looked for foundered after better seasons and the widespread deployment of irrigation restored the interior, only to flicker back to life when the limited carrying capacity of the land was once again tested in drought. In a report to the Willandra Lakes Region World Heritage Plan of Management Committee in the mid-1990s, Jim Bowler and John Magee declared that deeper understanding of the long histories of the landforms might teach that 'delicate balance, so necessary in establishing management practices' in the region. They noted dryly that one of 'the few reliable predictors of climate' in the region 'is the certainty that a run of good years will be followed eventually' by drought.

Russell Drysdale's painting 'Walls of China' is a canonical evocation of drought in Australia's arid interior. Painted during 1945, based on sketches the artist made during the journey with Newman, it was bought for £200 by the Art Gallery of New South Wales in late 1945 by the new Director, Hal Missingham. Missingham shocked the trustees with his expensive, eccentric purchase. They in turn 'reacted by immediately cancelling my powers, some claiming that tourists would be outraged at this stark portrayal of our land'.[25] The painting depicts the eroded Mungo lunette as a series of low, carved, red hills, the backdrop to an arid plain with a tortured, skeletal tree trunk, and a red, dusty sky above. Drysdale named the painting 'Walls of China, Gol Gol', although by 1945, the gigantic pastoral station had been subdivided, and the Walls spread across the smaller Mungo and Joulni Stations.

The reaction of the trustees suggests Lake Mungo's public beginning as a national landscape, before it had a name, was a desolate land of skeletons. Missingham's gamble paid off. The painting was exhibited overseas and was for many years one of few Australian paintings available to purchase as a colour reproduction.[26] It thus fed an idea of a landscape of ghosts and skeletons that was hostile, broken and desolate, but could also be integrated into a national psyche that located pride in the hardiness of the men and women who survived life on the

Figure 35: Lake Mungo presents a 'moving poem of destruction'. *(Photograph Michael Main)*

'long tragedy-track over scorched earth', as Newman described the blighted region on his journey with Drysdale. He noted of the country between the Darling and the Lachlan Rivers that 'however far from the great cities', and notwithstanding its desolation, 'this land was part of our heritage and our future'.[27] Otherwise, as Bonyhady suggested, the Walls 'remained almost unknown', and the ovoid saltbush plain at their base remained un-named until Jim Bowler's intervention in the 1960s thrust the region back into a national spotlight.

Characterised as poor grazing country, badly watered, semiarid, fringe, unforgiving, degraded, the lakes landscapes nonetheless began to be redeemed by the 'unparalleled record' of Quaternary climate change revealed in their sediments. They remained marginal landscapes. Then Jim Bowler's discovery of human remains in 1968, coming as it did at a time when some Australian archaeologists had begun to insist on a Pleistocene antiquity for Aboriginal Australians, catapulted Lake Mungo and the rest of the Willandra Lakes into the global archaeological canon. The cultural significance attached to the human remains attracted a World Heritage nomination and eventual listing. Without the concomitant publicity, the region could still be a geomorphologic footnote, eroding with every storm, plagued by goats and emus. As a World Heritage Area, it offers instead global standing, national significance, shared heritage and an incomparable marketing opportunity. Or as Bowler put it in 1998, as 'one of the best examples of ancient people–land interaction on time scales rarely matched anywhere in the world', the Willandra Lakes provide 'an opportunity to integrate physical environmental history with ancient cultural heritage, an issue of special importance to all Australians'.[28]

Chapter 9

HUMAN ANTIQUITY AND THE SCRIPTURE OF THE LANDSCAPE

Berriait or Barindji

Little has been written about the people who lived in the Willandra Lakes region at the time of and immediately prior to European settlement. Their history must be extrapolated from the limited observations of early settlers and explorers who, like Sturt and Mitchell, recorded their encounters with and speculations about the people they encountered nearby. A.L.P. Cameron, a local landholder, spent time with the people of the Riverine region between the Murray, Darling, Lachlan and Murrumbidgee Rivers in the second half of the nineteenth century. He identified the Berriait or Barindji group as the people inland of the rivers. Cameron wrote that 'bari' or 'beri' means scrub or rough country, although Norman Tindale recorded the Barindji as the 'people of the trees'. In 1904, A.W. Howitt wrote of these groups and their land, 'a large extent of country, without any permanent water, between the Darling, Murray, and Murrumbidgee Rivers', that, when surface water failed, they

> obtained a supply from the Mallee, a species of Eucalypt, and from one of the Hakeas. At times of drought they were forced to go to the rivers for water, and as these were occupied by other tribes such as the Barkinji and the Wonghibon, they had to fight their way in strong parties.

Cameron explained that they had a powerful sense of their separate identity from the river people, despite close similarities in language between the Berriait and the Barkindji of the Murray–Darling system: 'I suggested to my informants that they were part of the same tribe [as the Barkindji], but they would not hear of it.' He noted that their method for extracting water from hakea and mallee sustained them for four to five months every year. They had mostly moved or been removed from the Willandra Lakes area by the time Gol Gol Station was subdivided and the Mungo and Zanci leaseholds were established for soldier settlers in 1922.[1]

In his 1972 doctoral thesis on man and land in the Darling basin, Harry Allen, then a student in prehistory at the ANU, explained that funerary rites on the Murray River as reported in the nineteenth century shared similarities with those of the exhumed Mungo remains, suggesting an unprecedented degree of cultural continuity over time. In contrast, Thomas Mitchell recorded details of burial sites along the Lachlan, Bogan, Murrumbidgee, Murray and Darling Rivers in 1835 and 1836, noting much variety, but he reported no sign of cremation. The problematic relations between Mitchell's party and a group of Barindji in 1835 and 1836 probably led to the confrontation in May 1836 in which his men killed a number of Barkindji people, whom they mistook for the 'spitting tribe' (of Barindji) who had harried their passage through the region. Allen arranged the Barkindji group into three historical divisions. He included the Barindji in the East Darling division.[2] Allen also compared dietary habits and technology across the Murray–Darling region from 20 000 years ago to the present, as evidenced by the archaeological record and by nineteenth-century ethnographic accounts. He documented many other apparent continuities through time, though he also showed evidence of cultural change, in technological and dietary adaptations to the shifting ecologies of the terminal Pleistocene to Holocene. These included a diminished reliance on freshwater harvests and increasing use of seed and grain, as demonstrated by the appearance of grinding stones around 15 000 years ago.

The Lake Mungo archaeological assemblage includes a silcrete quarry on the western side of the lake, used to make many of the tools commonly found on the lunette and lake floor. The silcrete is also exposed along the shores of Lakes Chibnalwood and Leaghur, and is associated at both sites with toolmaking debris, including stone cores weighing up to a kilogram. Tools made from the silcrete are also found across the other lakes. Up to two kilograms of red ochre, which is not found within 100 kilometres of the lakes, coated the corpse of Mungo Man and now stains the bones. It must have been carried to the area or traded, and suggests that the Willandra people of 45 000 years ago regarded death rituals seriously, hoarding the pigment, but not stinting on the dead.[3]

At times, the shores of Lake Mungo were rich in resources before the last glacial maximum. Marsupials provided protein, skins, and bones to make fish spears or needles. From the lakes people gathered waterbirds, Murray cod, golden perch, crayfish, the freshwater mussel *Velesunio ambiguus*, and fibre from reeds for nets or ties. They found a ready source of stone at the silcrete quarry. Dunefields and grassland beyond the lake perhaps provided seed and grain. The lunette was probably lightly wooded, useful for shade and shelter, sap and toolmaking. Archaeologists John Mulvaney and Johan Kamminga envisaged 'well-fed, healthy people who had a varied high-protein diet, probably provided by a wide range of food resources'. They noted that historically, Aboriginal people of the Murray–Darling system collected freshwater shellfish in summertime. Analogy with nineteenth and twentieth-century Murray River people suggests that emu eggs might have been collected in late winter to spring, indicating trans-seasonal use of the lake area. Most of the marsupials represented by skeletal remains in middens, hearths and earth ovens were common in the region a century ago. Many prefer a woodland fringe habitat, but other species indicate more arid ecologies, suggesting a broad subsistence base and wide foraging and hunting by the people.

Like other historical sciences, archaeology is inexact and relies on narrative licence. Mulvaney and Kamminga cautioned that reconstructions 'telescope data across thousands of years into one day'. Such is the nature of the evidence that it is impossible to resolve whether the lakes were used continuously by one group of people or sporadically by different groups, whether occupation was cyclical, permanent, seasonal or an opportunistic mix. It is also difficult to determine if the shell lenses, midden remains and hearths represent single feasts or gradual accumulations. As Allen and co-authors observed in 2008, 'proximity in space does not mean closeness in time'. Adjacent sites and features can be separated by several centuries, or even millennia. The stone technology of the lakes region, famously designated the Australian core tool and scraper tradition by Allen and Rhys Jones in the early 1970s, thus reveals very little about population density, inventiveness, or indeed the variety of less durable material culture that might have been used (Figure 36).[4]

A number of workers, notably Allen himself as recently as 2009, caution that direct ethnographic analogy from living or historical Aboriginal people to the fossil record, and *vice versa*, has limited use. Colonial ideology and the fact of European invasion itself influenced the way Aboriginal people have been portrayed, and have portrayed themselves. The archaeologist Sarah Colley pointed out in 2002 that the profound impact of European explorers and colonial settlement on Indigenous societies, often ahead of physical contact, means that 'historical descriptions of Indigenous life may not accurately reflect what life was like' before British invasion. This seems something of an understatement. Consequently, workers have tended to 'extrapolate people's behaviour prior to AD 1788 in many different parts of Australia from information about better documented Indigenous societies in the north, west and inland regions'. Furthermore, prehistoric material culture is opaque. Particular objects might have had very specific, culturally and climatically determined uses which to a modern interpreter may appear counter-intuitive. In an archaeological echo of Henry Gee's caution to palaeontologists, Allen and his co-authors pointed out in 2008 that for over a century workers have assumed that Pleistocene Australians 'acted in ways similar to those observed by nineteenth-century ethnographers' and further assumed that 'archaeological evidence can be analysed in a cumulative fashion, fitted together like a jigsaw, to provide a composite model of human utilisation of the landscape.' Extrapolating from this tendency, Allen observed that indications of cultural continuity in areas like the Willandra Lakes region are 'as much a response to the way the data is brought together as it is to an archaeological reality', concluding that 'the majority of the conclusions which have been drawn from the archaeology of the Willandra Lakes owe their origin to the frameworks of interpretation used to draw the data together rather than to the archaeological data we have to hand.' General questions about Aboriginal demography in the late nineteenth century, stemming in part from the abundance of discarded stone objects in the desert, were put into perspective by J.W. Gregory in 1906, who reflected in *The Dead Heart* that 'The fact is probably due to the stone-using people having had no pockets.'[5]

In *The Custodians*, Jose shifted the cunningly disguised 'Lake Moorna' to west of the NSW border (even in an account of national identity, parochialism with its roots in pre-Federation protonationalism plays a part). He evoked Lake Moorna with an archaeologically informed

Figure 36: Bone point. *(Photograph K. Douglas)*

Figure 37: Walls of China, Lake Mungo looking north. *(Photograph Michael Main)*

sense of majesty, age and mystery, but one that was intrinsic to the landscape, independent of the works of humanity. Geology provided meaning, a new way of 'seeing' a landscape in the fourth dimension. Jose's geomorphologist Fritz Vogel (a sort of Teutonised Jim Bowler) 'had seen through the surface of the land with his X-ray eyes', revealing that there, 'in what was now a flat, parched, barren landscape, people could have lived a pleasant lakeside existence abundant in fish, flesh and fowl' (Figure 37). In Jose's recreation, and analogous to Bowler's encounter, the mystical experience was vindicated, not created by Moorna/Mungo Woman's discovery:

> without breaking into features of even the simplest narrative, the country went on for miles ... The dry sunken bed of Lake Moorna ... was the dominant feature only because it was even more minimal than the surrounding plain ... The negation of landscape as conceived in the pictorial tradition: no alps, no crags, no ruined towers, no nestling hamlets; in that lay its teasing eloquence.[6]

On 2 June 1836, Mitchell surveyed east and west of the Darling, not far from its junction with the Murray at the present site of Wentworth, and saw only 'unvaried desert features'. Drysdale's 1945 series of paintings of the region has been described as 'amongst the most terrifying images of devastation ever painted'. These days, the Walls of China appears on stamps, and visitors can pay $60 per person for a day trip to the National Park with Harry Nanya Tours. The deep past provides a new aesthetic, according to which the sparse, worn, dry, goat-infested landscape is reconfigured as an 'eerie, yet beautiful, moonscape', 'a constant source of mystery and delight', spectacular, surreal and dramatic. Dawn light 'on the pale ramparts of the Walls of China' is 'nature's theatre at its subtle best' and 'the feeling of serenity and sheer space is absolute'. Tourist destination, World Heritage property, geomorphological text book, archaeological wonderland, National Park, Aboriginal sacred site, contested country: the lakes landscapes, especially the Walls of China lunette, are now unambiguously national landscapes.[7]

Nineteenth-century naturalists and human antiquity

It is easy to forget that human prehistory had to be discovered.[8] Debate about the antiquity of man occurred against the background of the invention of glaciology and its incorporation into Lyell's uniformitarian paradigm, and speculation about extinction and the meaning of fossil fauna. As Donald Grayson and Martin Rudwick have shown, this bespoke intellectual landscape incorporated the signs of humanity in the form of tools and skeletons, eventually to convince mainstream naturalists that the evidence for a deep human antiquity was overwhelming.

Throughout the early years of the nineteenth century, claims for the discovery of human remains or material culture in association with extinct fauna and ancient landscapes were eagerly rejected. Many commentators were disturbed by the suggestion that human beings could be contemporaneous with the faunal products of earlier and 'lower' creations. Charles Lyell, for instance, considered that the dignity of man was 'at stake' and Charles Darwin was exercised by the implications for human origins of his ideas about natural selection. Eventually the discovery in western Europe of tools and human skeletal material in proximity to the fossils of extinct animals, most notably by the French customs director at Abbeville, Jacques Boucher de Crèvecoeur de Perthes, in the Somme valley gravels, convinced many European commentators of the contemporaneity of mankind and extinct fauna, although they baulked at his Mosaic framework. In Britain, mainstream scientific acceptance came more slowly. It was precipitated by an excavation at Brixham Cave, near Torquay, where in 1858–59 William Pengelly and Hugh Falconer discovered a clear association of flint tools with extinct fauna. Older cave evidence, previously rejected, was re-examined. Lyell's capitulation on the question of human antiquity appeared as a paper delivered to the BAAS at its 1859 meeting, 'On the occurrence of works of human art in post-Pliocene deposits'.[9]

The development of the discipline of archaeology from its antiquarian roots has been dealt with in some detail by a number of authors. Central to this shift from antiquarianism to the beginnings of a professional archaeology, and of critical importance later to the interpretation of Aboriginal material culture in Australia, was the idea, first clearly articulated in Richard Colt Hoare's work *Ancient Wiltshire* (1810–1821), that artefacts are more than simply objects of value in their own right. *Ancient Wiltshire* demonstrated that the material remains of the past are critical sources of information about the ancient people who left them. Excavations took on greater significance, not just as a means to uncover the buried treasures of hearth and barrow, but when undertaken scientifically, as a way to address specific problems of antiquity.[10]

Following from this focus on the historical information encoded in artefacts, the invention of the three-age system in Scandinavia revolutionised the practice of prehistory in the 1830s. Early in the century, a Danish historian, Vedel Simonsen, described three stages in the history of humankind, delineated by changes in material technology unearthed by antiquarians: a Stone Age, succeeded by a Brass or Bronze Age, in turn succeeded by an Iron Age. This system was adopted by Christian Jurgensen Thomsen, the curator of the National Museum of Copenhagen, who began in 1836 to rearrange his collections according to Simonsen's classification. Thomsen's pupil, John Jacob Rasmussen Worsaae, expanded the three-age system, unearthing further evidence of the sequence in the peat bogs of Denmark. The English translation of Worsaae's treatise, *The Primeval Antiquities of Denmark*, appeared in 1849, where it did not initially find wide acceptance. Commentators dismissed it as 'specious' and 'without foundation'.[11]

Eventually a number of English antiquaries accepted the fundamental applicability of the system, adding refinements of their own. It established a relative chronology which allowed systematic classification of tools by type and time, and for 60 years provided the central organising framework for British prehistorians. Most effectively, John Lubbock, Lord Avebury, proposed in a series of articles in the *Natural History Review* (expanded into his book, *Prehistoric Times*, in 1865) a division of Thomsen's Stone Age into a Palaeolithic Age 'when men shared possession of Europe with the Mammoth, the Cave Bear, the Woolly-haired Rhinoceros and other extinct animals', and a Neolithic Age, 'the later or polished Stone Age'. Lubbock's book went into five editions in its first 25 years. His work also retained the Bronze and Iron Ages of Simonsen, Thomsen and Worsaae. The term 'mesolithic' was introduced in 1866, expanding the system to five periods or ages. Australia was a prehistorian's candy shop – full of stone tools and the skeletons of the people who made them. Accordingly, living Aboriginal people could be slotted into a continuum of primitiveness according to the stone tool typologies of Europe, and craniometric analyses of the living and the dead.[12]

The age and origins of the Australians

It was against this background of an emerging science of race in Europe and Britain – which the discovery of a deep history of humankind altered profoundly – that metropolitan naturalists refined their speculations on the origins of the Australians and Tasmanians. But from their first encounters in the seventeenth century, British voyagers and Aboriginal people have provided each other with vehicles for representing themselves in contrast to strangers.

There are no written records of the earliest Australian traditions about the arrival of European men, but for the Europeans, the people native to the southern continent slotted easily, if problematically, into reflections about civilisation, degeneration and racial type, and metropolitan debates about the origins and spread of mankind. From William Dampier's first famous encounter in 1688 to James Cook's, nearly a century later, English impressions were diverse, but highly codified, loaded with signifiers of savagery or civilisation, and influenced intractably by the behaviour of the people being described.[13] Dampier declared of the people he saw on the coast of Western Australia that:

> The Hodmadods ['Hottentots' of the Cape of Good Hope], though a nasty People, yet for Wealth are Gentlemen to these ... And setting aside their Humane Shape, they differ but little from Brutes ... They have great Bottle Noses, pretty full Lips, and wide Mouths. The two Fore-teeth of their Upper jaw are wanting ... Neither have they any Beards. They are long visaged, and of a very unpleasing Aspect, having no one graceful Feature in their Faces. Their Hair is black, short and curl'd, like that of the Negroes; and not long and lank like the common Indians. The colour of their Skins, both of their Faces and the rest of their Body, is coal black, like that of the Negroes of Guinea.[14]

Eleven years later, he was no kinder, having determined that 'they all of them have the most unpleasant Looks and the worst Features of any People that ever I saw, though I have seen great variety of Savages ... black Skins, and Hair frizled, tall and thin'.[15]

In comparison and self-conscious contradistinction, Cook's botanist on the *Endeavour*, Joseph Banks wrote of people he observed on the Endeavour River on 12 July 1770 that 'certain it is that Dampier either was mistaken very much in his account or else that he saw a very different race of people from those we have seen'. Banks and Cook stressed the un-likeness of the people of the east coast either to Dampier's account, or to 'negroes' more generally, instead seeing comparisons with Pacific islanders. As settlement proceeded, interest in Indigenous Australians did not fade with greater familiarity. Well into the twentieth century, scarcely would an inland explorer complete the public account of his expedition, than he would supplement it with a commentary on the Australian native, his origins, nature, manners, customs and future, and reasons for his decline.

Well before Captain Arthur Phillip and the First Fleet made landfall at Sydney Cove in January 1788, Europeans speculated about the age, origins and closest cultural affiliations of the Australians. They were co-opted into a scheme of natural history that saw Australia as a land of living fossils, its people static and primitive like its flora and fauna. This attitude had consequences for the professional development of archaeology in Australia. If Aboriginal people had no deep past, what was the point in studying it? From the time of Ranken's and Mitchell's establishment of the *longue durée* of Australian marsupials in 1830, naturalists were eager to establish or refute Aboriginal synchronicity with the bone beds at the Wellington Caves. At Lancefield Swamp and western Victoria's Lake Colongulac in the 1840s, excavators claimed that Aboriginal tools were associated with the fossils of giant marsupials, and human remains of questionable antiquity were regularly exhumed on the Darling Downs in Queensland. Naturalists commonly associated Indigenous accounts of now-vanished or unidentified beasts with the fossil remains of extinct megafauna at these and other locations. The problem was partly one of taphonomy. As witnessed at Lake Mungo a century later, the deep past in Australia is often literally very shallow (see Figure 38 below). Depth of artefacts or bones could not be correlated with age.

However, the inconclusive nature of the evidence for human occupation at the best-known fossil sites was no deterrent to metropolitan scientists eager to connect Australian evidence to the better-known Northern Hemisphere data, and untangle the chronological, possibly causal, relationships between prehistoric humanity, the glacial epoch and the extinction of the megafauna. Geologists and palaeontologists were responsible for much of this work in the nineteenth and early twentieth century; for example, T.W. Edgeworth David, Robert Etheridge Jnr, Robert Lockhart Jack, Gerard Krefft, John Walter Gregory and Alfred William Howitt, all combined geological work with ethnological speculation and *vice versa*.

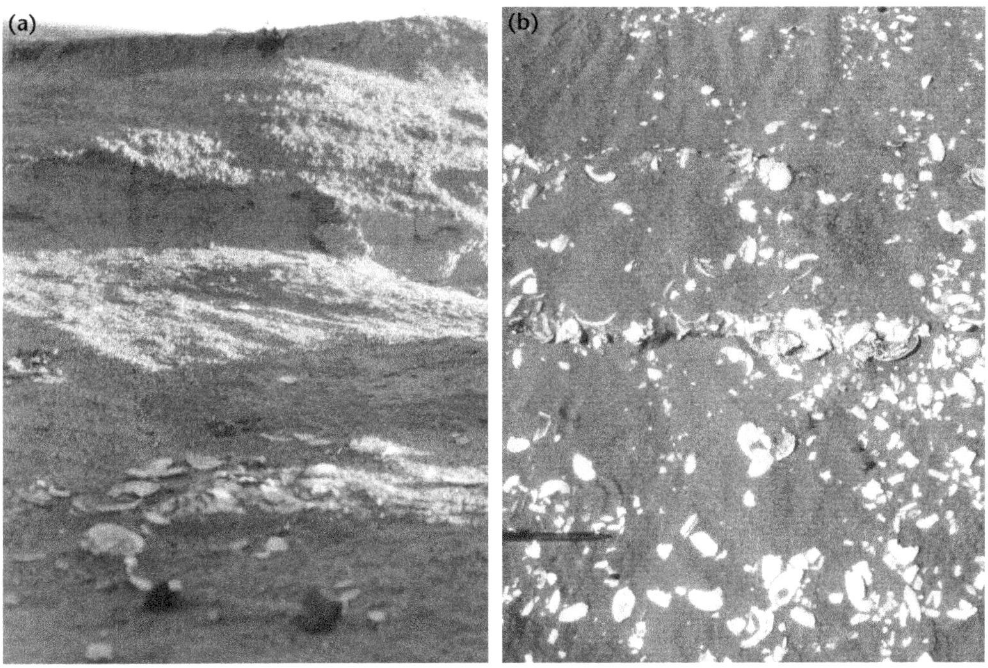

Figure 38: Ancient *Velesunio ambiguus* midden, Lake Mungo (38a as seen in the wider landscape and 38b a close-up view). *(Photographs K. Douglas)*

The absence of stone tools in direct, indisputable association with megafaunal remains made correlation with Northern Hemisphere ice age cave sites more difficult. Studies of Aboriginal people were considered very important to establishing universal racial classifications. Their origins, antiquity and affiliations could be fixed by looking at their bodies, their skulls, faces and hair, their languages, material technology and their arts, as well as their oral traditions. Before the establishment of radiocarbon dating methods in the 1950s, it was very difficult for researchers to determine the age of artefacts and sites. Like Australian landscapes, Australian soils and Australian flora and fauna, Aboriginal stone tools did not easily fit European typologies. The five-age system was useful only for placing Aboriginal and Tasmanian people in a primitive stasis. For example, in 1894 the renowned anthropologist E.B. Tylor claimed for the Tasmanians 'a stage lower than that of the Quaternary period' in Europe. The 'lowest of normal tribes', he described them as 'living representatives of the early Stone Age, left behind in industrial development even by the ancient tribes of the Somme and the Ouse'. He encouraged study of the Tasmanians as 'a subject materially affecting the history of the Stone Age, namely, the persistence among these modern savages of a state of stone implement making comparable to that of mankind in their remotest acknowledged antiquity'.[16] Eventually developments in phrenology and physical anthropology (themselves reliant on a trade in body parts) facilitated the assignation of racial types. A letter written in July 1847 from Ronald Gunn to Hobson, then resident in the Port Phillip district, highlights the nonchalance with which collectors gathered the skeletal remains of Aboriginal people into museums and private collections. Gunn casually explained to his friend that he was 'off to Hobart Town about the day after tomorrow, but only for a few days. I shall pick you up some skulls and bones on my return Home – also fossils.'[17]

In 1884 or 1885, William Naish, a stockman, found the Talgai cranium in a dry creek bed near Warwick in south-east Queensland. The skull, apparently either heavily deformed or 'primitive', belonged to a young boy, and was found in sediments that suggested a certain antiquity. Despite Naish's discovery, in 1892 Jack and Etheridge asked if man had a geological history in Queensland and answered no. In contrast, citing physiological, linguistic and artistic evidence, Howitt ruled in 1904 that the Australians had been here a long time: 'the conclusions to which I have been led as to the origin of the Tasmanians and Australians necessarily demand a vast antiquity on the Australian continent for the former, and a very long period of at least prehistoric time for the latter.' But the question was by no means closed. Naish showed Edgeworth David the Talgai site in 1914, and David spent the next decade trying to prove Aboriginal antiquity on the basis of the find. In 1925 the debate played out in newspapers, David decrying those commentators who judge antiquity by depth rather than 'primitive characteristics'. A letter from Miss Georgina King on 25 May 1925 in the *World's News* asserted that she had heard that the Talgai cranium was a skull from the Darkey Flat Massacre. Geologist and radiocarbon dating pioneer Edmund Gill revisited the site and the issue in the 1960s. In the 1970s the skull was dated by radiocarbon to 13 000 years, but by then the debate had moved on, largely due to the greater antiquity of the Lake Mungo discoveries.[18]

In 1929, Norman Tindale and Herbert Hale, both then at the South Australian Museum, undertook the first systematically documented, stratigraphically meticulous archaeological excavation in Australia at the Devon Downs rock shelter in South Australia. In 1936, Fred McCarthy of the Australian Museum likewise conducted a systematic excavation at the Lapstone Creek rock shelter in the Blue Mountains. The excavators recorded change over time in stone tool technology at both sites, laying the groundwork for future archaeological syntheses. Apart from the work of Tindale, Hale and McCarthy, archaeology received very little institutional support over the next 20 years, but was the domain of amateur collectors and antiquarians.[19]

John Mulvaney is widely regarded as the founder of professional archaeology in Australia. Having studied archaeology at Cambridge, Mulvaney returned to the University of Melbourne and in the late 1950s taught the first course in prehistory at an Australian university. By 1964, prehistorians were employed at universities in Melbourne, Canberra, Sydney and Armidale. In 1960, Mulvaney and Bernie Joyce established radiocarbon dates of 14 000 years at Kenniff Cave rock shelter in central Queensland, pushing human occupation of Australia unequivocally back to the Pleistocene. Then in 1968, Bowler found Mungo Woman. He and his colleagues published a conservative date of 18–25 000 years in *Nature* in 1972, and the antiquity of the Australians assumed international significance in its own right. Within the decade, David Yencken, Secretary of the Department of the Environment during the Whitlam Government and the foundation chair of the Australian Heritage Commission, wrote and steered the World Heritage nomination of the Willandra Lakes region through the Australian Government to UNESCO.[20]

The Mungo remains assumed ever-greater importance when Alan Thorne contrasted them with the more 'robust' and 'primitive', but stratigraphically younger Kow Swamp remains, found on the Murray River near Echuca in the late 1960s. He devised an evolutionary scheme for Aboriginal Australians, suggesting that they were a hybrid of two waves of physically different Pleistocene colonisers. The first wave was a robust descendant of the Java *Homo erectus* whose descendants are exemplified by the Kow Swamp skeletons. The later wave of more delicately constructed Lake Mungo-type people arrived in Australia direct from southern China, independently descended from a north Asian *H. erectus* like Peking Man. Thorne believes that the younger Kow Swamp remains are a relict population of the older Javan type. Other workers claim that the robustness of the Kow Swamp remains (which have subsequently been re-dated to the last glacial maximum, at least 18 000 years ago) is simply the result of a community becoming isolated genetically due to the rigours of the ice age climate. Compelling evidence suggests the distinctive shape of the Kow Swamp skulls is a result of deliberate deformation by pressing on the foreheads of babies. Colin Pardoe suggested in 1995 that differences in skeletal morphology reflected the extension of the Pleistocene arid zone into much of the Murray Basin rather than the coexistence of two different species of human being. He claimed that the Willandra skeletons show a people 'typically arid-zone adapted': lean, with long limbs. Some workers even suggest that Mungo 'Man' is in fact an adolescent woman and that the so-called gracile–robust debate could be an artefact of sexual dimorphism.[21]

Thorne and Bowler have in the last decade been involved in separate redating of Mungo Man. In 1998, Bowler and D.M. Price published revised dates of 45 000 years for Mungo Man.[22] The next year at the Australian National University, Thorne's group published a new date, revolutionary for involving direct dating of the bone, claiming to push human occupation back to 61–81 000 years. It was first announced in a seminar at the ANU on 21 May 1999 with considerable media interest, then published in the *Journal of Human Evolution*.[23] Bowler and John Magee countered with 'a sceptic's view' in the journal the next year, pointing out correctly that the bones cannot be older than the grave sediments in which they are buried. Another paper by geochronologists Richard Gillespie and Bert Roberts also suggested the new date was too old. In the same issue, ANU dating specialists Rainer Grün and Nigel Spooner, together with Alan Thorne, published their detailed reply, suggesting that Bowler, Magee, Gillespie and Roberts had failed to understand their data, claiming that their own dating methods, directly from the bone, are 'more' definitive, and lamenting that 'we are still not able to read the minds of our colleagues'.[24] Bowler responded in an opinion piece in the *Australian* in 2001.[25]

Bowler attributes the controversy about the age of the skeletal material to intellectual demarcation disputes and rivalry, personality clashes and factionalism between and among the disparate disciplines and institutions involved in the studies of human antiquity. He also takes some responsibility for other scientists' misunderstanding the former shape of the Willandra Lakes landscapes. His idealised models of lunette stratigraphy in the lakes are canonical images in texts dealing with archaeology and geomorphology in the region. They depict the ancient dune profiles as smooth layers overlying each other, leading to the assumption that the characteristic yardang-style erosion and gullying in the region is an artefact of pastoralism (for example, see Figure 33 on page 142). However, the dune stratigraphy is far from a simple layer cake (see Figure 39). Thorne and his colleagues' failure to recognise that the grave of WLH 3 was dug into ancient gully infill might have resulted in an artificially older date for Mungo Man.[26]

The age of the Mungo remains has significance for questions about the extinction of Australian megafauna. Several models have been proposed to account for the disappearance of most animals of 100 kilograms or more from the continent during the late Pleistocene epoch. Whether climate or ecological change at any of a number of scales, or one of the various models requiring human intervention, each theory relies on establishing the earliest human habitation of the continent, as well as the timing and extent of climate shifts, and the last appearance of megafaunal species. A few *Procoptodon*, *Zygomaturus* and *Macropus titan* remains (respectively, a gigantic, browsing, flat-faced sthenurine kangaroo, a hippopotamus-sized grazing quadruped with a mysterious nose and a huge grazing kangaroo morphologically similar to the modern grey) have been found in the Willandra Lakes, but not in any clear association with human remains. But the dates of 45–50 000 years for the earliest human skeletons in the area fall within the most recent widely accepted dates for the youngest remains of extinct megafauna on the continent.[27]

The material remains of the past

> At issue are fundamental questions of ideology and ownership. Does the history of humans in Australia, which is now known to date back 30 000 years, at least, belong to the ethnic descendants of those first inhabitants? ... Or is there some wider claim, of science and common human concern, to rights of access to relics of the past? Is it, perhaps, pushing an ethnic group's rights over its own past a long way to suggest the skulls of Lake Mungo, which are now upwards of 1000 generations old, 'belong' only to their assumed modern descendants?[28]

As the above extract from archaeologist Christopher Chippindale's 1985 article in *New Scientist* suggests, archaeologists in Australia and overseas have been forced over the last three decades to engage with newly vocal Indigenous stakeholders and newly recognised obligations other than those required by scientific enquiry. The past, its material remains and their interpretation have become contested, unstable ground. Today the repatriation debate is a major preoccupation of museum curators, professional and academic archaeologists and anthropologists. It is part of the cultural, ethical and legislative landscape of collection, curating, analysis and heritage which archaeologists cannot afford to ignore. It is itself a multiply contested, emotive area, which defies generalisation.[29]

In 2002, archaeologists Jane Hubert and Cressida Fforde wrote of repatriation that it is a way in which people can affirm 'their pre-eminent right to make their own decisions regarding what should happen to their ancestors' remains. In this way they can lay claim to their own pasts and determine what should or should not be part of their cultural heritage.' Particularly in colonised countries and for people who have been dispossessed of their land and material culture, 'the issue of the remains of the dead has acquired an added significance', because of the historical 'plunder' of 'what is now called [their] "cultural heritage"'. Of most current significance,

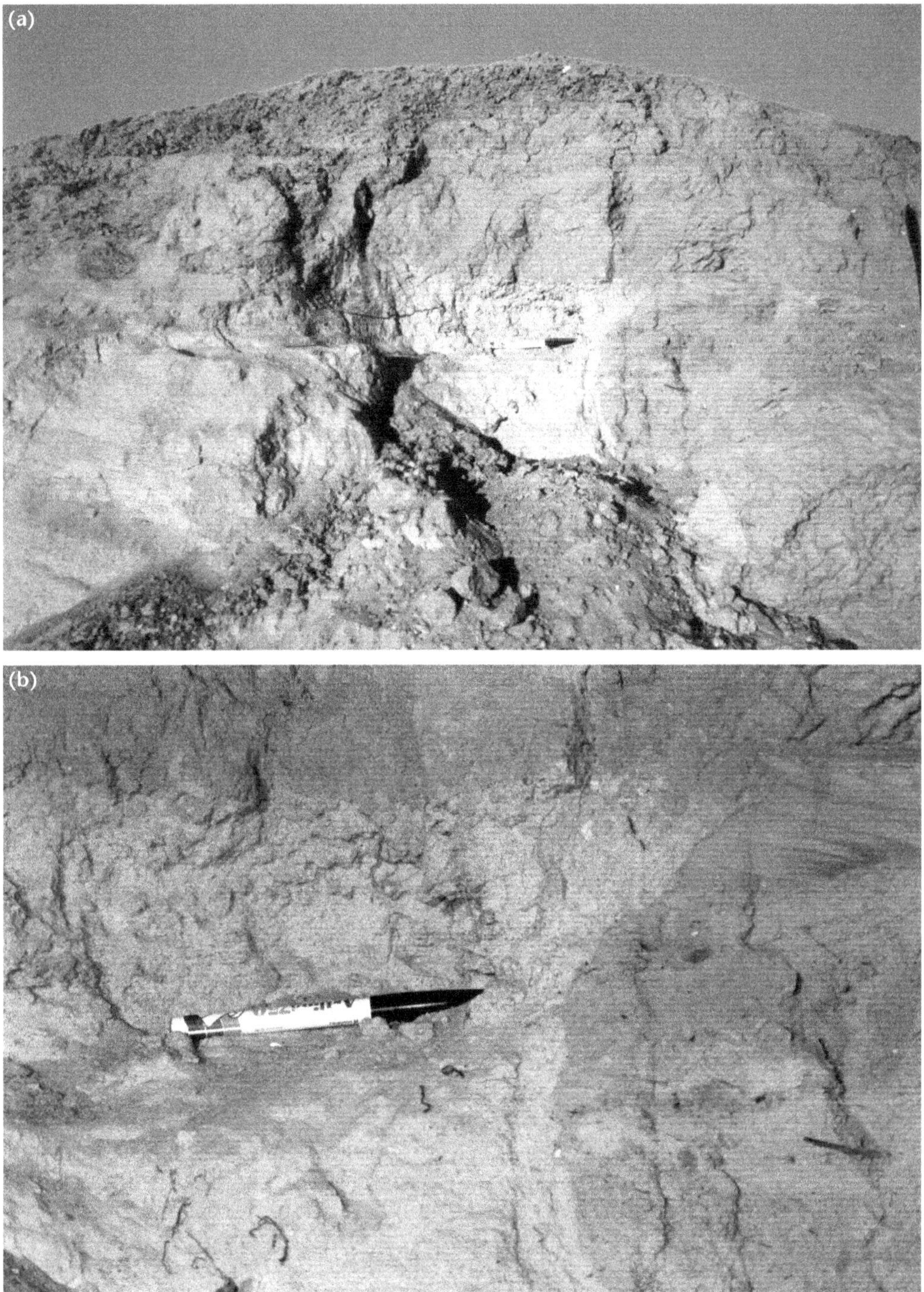

Figure 39: Gully at Lake Mulurulu showing former topography in the form of palaeogully fill and palaeo-yardang (39a shows the infill at a distance and 39b a close-up view). Lighter sediments to left of image are younger than darker sediments to right. Cross bedding evident in sediments to right of image. *(Photograph K. Douglas)*

'among the many things that were taken back to Europe were skeletons (especially skulls), mummified bodies, limbs, shrunken heads and various other anatomical specimens ... Human remains have also been removed by archaeological excavation, to be studied by archaeologists and physical anthropologists, and to be stored and displayed in museums and university departments.' In this environment, the context of the Mungo remains as the oldest in Australia assumes even greater significance.[30]

Material remains of the past are cultural artefacts with powerful political resonance. They are part of the fabric for the construction of an illusory national unity, the glue that sticks us to the land. The rhetorical confusion of representation of the repatriation debate and archaeology in the popular media locates 'knowledge' or 'the past' – rather than the possibility of *knowledge about the past* – in bones and cultural material. For instance, in a popular archaeology textbook, eminent prehistorians Colin Renfrew and Paul Bahn pointed out that 'The past is big business – in tourism and in the auction rooms. The past is politically highly charged, ideologically powerful and significant. And the past, or what remains of it, is subject to increasing destruction.' 'The past' is not a thing to be bought or sold, destroyed or salvaged. What Renfrew and Bahn intended to suggest is that its material remains are endangered. Knowledge, history and power are understood to reside in objects. The Mungo remains are reconfigured as Adam and Eve, symbolising the professional beginnings of a discipline and the backward extension of a nation as well as the spiritual heart of an Indigenous population and perhaps a new beginning for academic–Aboriginal reconciliation.[31]

Archaeologist Isabel McBryde has asked whether her discipline has 'taken possession of a past that is not ours, and either unwittingly or paternally collaborated in its appropriation as symbol of national identity'? She suggested that Australians consider why Australia's nominations for cultural sites for World Heritage listing 'have all been places of outstanding Aboriginal cultural heritage'. (This situation has changed with the World Heritage listing of Melbourne's Royal Exhibition Building and Exhibition Gardens in 2002 and the Sydney Opera House in 2006.) She then questioned whether this constitutes 'recognition? appropriation? or relegation of a dynamic culture and its creators to be seen once more as components of natural history', in a rehearsal of nineteenth and early twentieth-century racial science orthodoxies. It is perhaps as much a consequence of Indigenous Australian holistic understanding of what European Australians might call landscape or natural history, of the expression and interpretation of World Heritage criteria, and of UNESCO's introduction of a category of 'cultural landscape' to the World Heritage list as it is a sign of the intentional or insensible appropriation of Indigenous sacred places and objects by governments, archaeologists and heritage bureaucrats. Many archaeologists, McBryde included, work collaboratively with Indigenous communities, and increasingly Aboriginal people train as professional archaeologists. And McBryde perhaps overstated the connection between scientific understanding of Aboriginal origins during the nineteenth century and heritage practice in the late twentieth century. Nevertheless, as McBryde showed, the rhetorical link between Aboriginal antiquity and national patrimony is powerful and seductive.[32]

As Chippindale hinted, some archaeologists argue that the great age of human remains like those from Lake Mungo transcend race and ownership: it is impossible to trace modern affiliations or to gauge the wishes of the long dead, and their impact may be decisive in debates about the origin and radiation of modern human beings, so they should be available to all people in the world. This is justification by globalisation. John Mulvaney has long been an impassioned advocate for the 'protection of cultural heritage as "a national possession" essential to collective self-perception'. But in Mulvaney's view, archaeology is a transnational resource. As Renfrew and Bahn expressed it, 'The world archaeology is something in which we can all share.'[33] Except for the canonical Mungo Man and Mungo Lady, workers refer to the other Willandra Lakes remains for the most part as the impersonal 'Willandra Lakes hominids', nomenclature which works to distance the burials from an emotive, personal, political, human present. Many Indigenous Australians and their advocates (including archaeologists) argue that the remains are the ancestors of the traditional owners of the region and therefore constitute Aboriginal cultural property and should be 'returned' to the Indigenous 'owners' for appropriate disposition. We may have no access to the wishes of the interred or their families, but the very fact of their interment implies attitudes to burial and an afterlife suggestive of a desire to stay buried. On the one hand is the assumption of the 'universality of [archaeological] values'. On the other are the moral claims of dispossessed people to cultural property, and the assumption of a homogeneous community of Aboriginal meaning and culture through time.[34]

These debates are of course not restricted to Australian archaeology. The predicament of many North American archaeologists is characterised by situations like the acrimonious debate about the discovery, acquisition, investigation and disposition of Kennewick Man, found in eastern Washington State in 1996. Among the most complete ancient skeletons excavated in North America, its uncertain cultural and genetic relationships are balanced by questions about the nature of the truth claims of Western science versus those of indigenous traditions and the ownership of pre-Columbian material culture.[35]

Those contemporary prehistorians who are opposed to repatriation counter the arguments of McBryde, Hubert and Fforde by pointing out that their aims and methods are different from the physical anthropologists and comparative anatomists of the nineteenth and early twentieth centuries. They thereby fail to appreciate the legacy of anger and disempowerment stemming from the racial stereotyping and essentialism on which their disciplines are founded. They argue their right to study human remains on the basis that knowledge about human history should be global and that each of our pasts is all of our pasts. For example, in the Kennewick Man saga, investigative journalist Jeff Benedict and forensic anthropologist Doug Owsley asserted that, when the US Army Corps of Engineers, acting in the perceived interests of Native American tribes in Washington State, seized the Columbia River skeleton from archaeologists, it was threatening 'a wide body of knowledge about *our* past and *our* history that would be lost forever if the bones were reburied'. They argue from philanthropic and humanistic principles. As Renfrew and Bahn insisted, 'the essential truth' is 'that archaeology is an exciting quest – the quest for knowledge about *ourselves* and *our* human past'. Arguments are generalising, universal and scientistic, in counterpoint to the personal, spiritual and cultural arguments of Indigenous people and their representatives.[36]

As Rodney Dillon, then Chair of the Culture, Rights and Justice Board of the now defunct Aboriginal and Torres Strait Islander Commission indicated in an open letter to the UK Culture Secretary (published in *New Scientist* in 2004), it is not a simple case of rationalistic scientists and research institutions versus antiscientific Indigenous people and their representatives: 'Indigenous people are not backward or anti-science: we are aware of the possible benefits of research.' Dillon highlighted inconsistencies in British legislation: 'The Human Tissue Bill ... has made it clear that the scientific benefits from tissue and organs taken from deceased patients do not take precedence over the rights of those people and their families.' The same applies to organ donation in Australia. Why, as Dillon asked, should the rights of Indigenous people over the remains of people they identify as their ancestors 'be sacrificed for the alleged benefit of humanity in general when others are not asked to make such a sacrifice?' Without compromise, the controversy is irreconcilable.[37]

Philosophical issues attached to geological heritage are at first glance less thorny than those linked to the material remains of the human past. The questions of who owns the past (its physical remains) and who controls the past (perceptions and descriptions of the past) appear less sensitive when dealing with 280 million-year-old glacial deposits at Hallett Cove or the fossil traces of Precambrian metazoa at Ediacaran sites in South Australia, as opposed to the remains of somebody's meal or ancestor in the human past. So geological heritage operates very differently at the archaeological sites of the Willandra Lakes compared to Hallett Cove and Lake Callabonna.

Non-human material that may have informed Aboriginal accounts of their own pasts is more complex. For instance, nineteenth-century naturalists and their informants reported many Aboriginal accounts which linked megafaunal bone fields such as the Callabonna site with ancestral figures and the formation of the land and its features. In the dispute over Tyrannosaurus Sue, a huge and unusually complete *Tyrannosaurus rex* specimen, local Sioux claimed that the fossil was a mineral resource akin to oil and therefore belonged to the Native Americans who controlled the reserve, in the same way that oil leases do. In 1996, fossil dealers illegally removed several 120 million-year-old, 50-centimetre long dinosaur ichnofossils from Crab Creek, near Broome in Western Australia, sending 'shockwaves' through the palaeontological community. According to local Aboriginal people, the tracks have great cultural significance as marking the passage of a giant emu man. A posting by Chris Nedin, a South Australian palaeontologist, to an online forum at the Chicago Field Museum of Natural History, noted that the custodian, Joseph Roe, 'now fears spiritual retribution against him and/or his family' and that the thieves showed 'a callous disregard for the beliefs and customs of the local people'. Museum palaeontologist John Long noted that the trackways were sacred to local Rubibi community, and formed part of the Lurujarri Heritage Trail. Local elders cursed the perpetrators. Such events and the World Heritage recognition of Uluru and Katatjuta as a cultural landscape suggest there is no such thing as non-cultural heritage in country as saturated with human meaning as Australian landscapes.[38]

Repatriation and national heritage

Elsewhere in *The Custodians,* Ralph Kincaid, a Welsh archaeologist and thinly disguised proxy for the late Rhys Jones, argued that archaeology's 'real treasure was not gold or silver, but time itself'. Europeans 'wove themselves into the story as excavators and articulators' configuring Australian prehistory as 'an essential nation-forming discipline'.[39] Jose's evocation vindicates McBryde's and fellow archaeologist Denis Byrne's claims that Australians appropriate the signs of the human past as part of a project to build, validate and shore up collective national identity in the creation of a 'deep nation'.[40] Lake Mungo has become a place of awed contemplation for heritage tourists. The mystery and the monumentalism of the deep past evoke reverence at Mungo as at Callabonna and Hallett Cove, but the human past provides extra gloss (see Figure 40). Despite the site's World Heritage classification under both natural and cultural criteria, it is for illumination of the human past that the area is chiefly and popularly celebrated, even by geomorphologists, and the landscape is reconfigured as a place of wonder and beauty on the strength of its human associations. Bowler, for example, observed in 1998 that the presence of complex burial rituals (involving ochre and cremation) provides 'one of the dramatic mysteries of ancient human cultural development' and the death of an individual casts light on both the people who buried him and 'the Ice Age environments that supported them'.[41]

As Bowler and others have pointed out, this association can be problematic. Western science requires that knowledge be shared, notwithstanding the demands of institutional brinkmanship and intellectual copyright. Consequently antique human remains, even those that have already been studied, should be stored against the possibility of new insights into

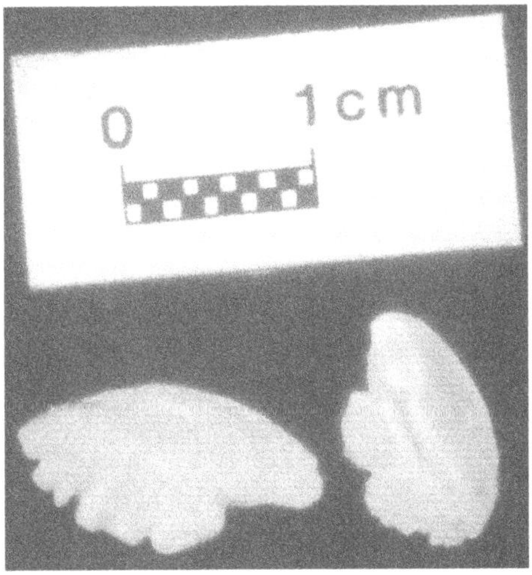

Figure 40: Fish otoliths in middens in the Willandra Lakes lunettes provide information about palaeoecology, human activity and diet. *(Photograph K. Douglas)*

human descent, demographics, diversity and disease. But people of Mutthi Mutthi, Paakantyi and Ngyiampaa descent claim custodianship of the land and human remains, severely limiting the freedom of scientific research. Such issues of the ownership or custody of material remains have not yet clouded non-archaeological localities such as Callabonna and Hallett Cove, except where they overlap with cultural or archaeological sites. However, geological and national heritage generally are perceived by some as being under siege by a politically driven, 'nonscientific', 'antiscientific' or 'creationist' Indigenous activism.[42]

The North American situation is clouded by the current debate about whether intelligent design (or creationism) can or should be taught in school biology classes as a legitimate alternative to evolution. When Native American representatives in northern Washington State insisted that 9000-year-old Kennewick Man, with his supposedly Caucasoid features, is their ultimate ancestor, and Federal legislation backed them up, alarms rang for scientists from many disciplines other than archaeology and anthropology. They publicly decried the evil 'philosophical relativists', troublesome 'Indian chauvinists' and 'Indian tribe creationists', dangerous 'anti-scientific dogma' and 'guilt-ridden' 'liberal intellectuals' who sought to limit intellectual freedom.[43] With rather more circumspection, John Mulvaney, writing about the 'return' of skeletons from Kow Swamp to the Echuca Aboriginal Cooperative in August 1990, described that outcome as 'a triumph of bureaucracy and irrationality over prudence and positive, collaborative racial relations', the replacement of 'white violence and repression with black intellectual totalitarianism'. He observed that 'it is not simply the Kow Swamp relics which are at stake, but the future of past Aboriginal culture, and the freedom of all peoples of any race to study it.'[44] He asserted that claims to absolute knowledge of the past, representing 'an amalgam of Dreaming beliefs and anti-evolutionary Christian fundamentalism', challenge the 'intellectual freedom of every Australian, especially future Kooris'.[45]

When someone's cultural property and ancestors are reconfigured as other people's geological heritage and scientific data, the purpose of a national or transnational treasure chest of places worth keeping is called into question. Is it to safeguard things and places for posterity (the notion of which is itself undermined by the fundamental geological precept of change through time), for 'all humankind', for 'science' and 'the future'? Is it a noble exercise fostered by philanthropic governments and the United Nations, which care about 'the past' and its material remains for their own sakes? Is it a cynical or an idealistic exercise in extending national unity into the pre-national past, in the interest of a cohesive nation state, by appropriating aspects of indigenous cultures and pasts arbitrarily regarded as having value? In suggesting that a nationalistic agenda may underlie the celebration and appropriation of antiquity in Australia, I do not mean to impute cynicism to those involved in heritage or archaeological studies, just as the suggestion, like Chippindale's or Benedict's, that Indigenous claims for reburial are entirely or largely politically motivated oversimplifies the situation. The reasons are multiplicit, entwined, mutually incompatible and often unfathomable.

Most Australian archaeologists, and their governing body, the Australian Archaeological Association, support the transfer of post-1788 remains to communities, 'because social and spiritual considerations outweigh other factors'.[46] Those who oppose reburial or destruction of 'prehistoric' human remains point out that Pleistocene remains are hundreds of generations removed from living people. Any line of descent from the Willandra Lakes remains is to Aboriginal Australians everywhere, rather than to those people now living in a specific region, like Balranald or Echuca on the Murray River. Important information about demography, disease, migration and descent (that could be understood as belonging to all Indigenous Australians, if not all Australians, including those as yet unborn) is lost when skeletal material is destroyed. Reburial can be even worse, because if it is not properly documented, the integrity of entire sites is compromised.[47] Mulvaney pointed out that 'controlled scholarly enquiry and racial collaboration, which synthesizes these finds and re-assessments with traditional Aboriginal knowledge, benefits all in a free society'.[48] There is a fear among some museum curators that repatriation of any cultural property will 'open the flood gates to the destruction of collections'.[49]

Those who oppose the study of Aboriginal skeletal remains irrespective of age observe that the age of an ancestor is irrelevant spiritually. They can also point to a history of grave-robbing, cultural looting, dispossession, and lack of consultation.[50] Furthermore, textbooks and articles have displayed photographs of the human remains. Initially this may have involved ignorance of the distress and offence the images would cause many Aboriginal people, but newspapers, journals and magazines continue to include these photographs, particularly the charismatic images of the partially excavated Mungo Man skeleton taken in 1974.[51]

Repatriation is often equated with destruction. At a conference in Melbourne in 2001, archaeologist Michael Pickering, the National Museum of Australia's Director of Repatriation, pointed out that successful repatriation is about empowerment, and need not even involve the physical transfer of property, simply an acknowledgement of custodianship or ownership, and an undertaking to consult. It can be seen as an act of good faith. The struggle accompanying the interpretation and disposition of the human past in Australian history and prehistory is partly about atonement, reconciliation and human dignity as much as it is about science, 'ideology and ownership'.[52]

This is not to suggest that the discoveries and publications of archaeologists over the last 35 years have provided no material benefit to Aboriginal people.[53] The phrase '40 000 years' entered the vernacular during the early 1970s as a direct response to the publication of Mungo dates.[54] The temporal expansion allowed by this archaeological work accompanied changing sociopolitical agendas of both scientists and Aboriginal people, as Jim Bowler pointed out in an address to the Royal Society of Victoria in October 1981, as the recipient of the society's Research Medal for the year, on the topic 'From Mungo to Noonkanbah: Scientific development or racist exploitation?'.[55] He worried that refusal by traditional groups to permit the study of Aboriginal prehistory could serve to propagate the implicit assertions of Eurocentric national history, that non-literate people have no history.

Bowler pointed to problems of a cult of personality in Australian academia, and to intellectual rivalries spanning at least three institutions – Melbourne, Sydney and the ANU – which have affected interpretation of the Willandra Lakes material, and further politicised an already volatile situation. This cult of personality applies to human remains and cultural property as well as to academics. Mungo Lady has given 'a hell of a lot' both to archaeologists and to Indigenous people, as Joyce Smith of the Mutthi Mutthi people said in 1999, calling for her reburial at the spot where Bowler found her: 'They just have to take this one more step in putting this very sacred person to rest. She has given so much to our people and she's given the scientific world a hell of a lot, too.'[56]

Bowler himself pointed out the inability of scientists in the Willandra Lakes Region World Heritage Area to manage the cultural and skeletal material constantly emerging from the dunes.[57] Mungo Lady's spiritual importance now outweighs her scientific significance. Other skeletons (and phenomena like the trackways of 2004) emerge and will continue to emerge out of the lunettes, lake shores and dunefields which can reveal even more than she has about Aboriginal origins and human adaptation to late Pleistocene environments. So why the reluctance to bury this one, pawed and pored over, over-determined bundle of burned bones from a calcrete block? It is what she represents, as the first marker of great human antiquity in Australia that makes her as important symbolically to archaeologists as she is demographically. Courtesy of her narrative charisma, the timing of her discovery and her media prominence, she has been gathered into the bosom of the nation to represent all our beginnings.

'The desired future'

typically Australian in character, of immense antiquity by any standards and abundant in their diversity. [The lakes] form a repository of physical and human history unrivalled anywhere in this land.[1]

'The desired future' of the title is found in the watered past.[2] Australians find both comfort and trepidation in our new awareness of larger climatic cycles beyond anthropogenic effects. The desired future also rests in a landscape that can accommodate intellectual endeavour, limited pastoralism, cultural tourism and the requirements of its Aboriginal custodians. It is an aspirational landscape as devoid of the 'clinical dissection' of other people's prehistory as it is of 'Indigenous Creationists'' grandstanding. It is a landscape where the overwrought past of bodysnatching, craniometrics and the hunt for an essential primitive can be re-interred peaceably, while those archaeologists and physical anthropologists who feel they are fighting for the viability of their discipline can negotiate a future wherein the study of human remains is welcomed by Aboriginal people, but conducted sensitively. Unfortunately, no knowledge is apolitical, and the past does not stay buried, as the Willandra Lakes reveal in vivid cameos. Lake Mungo is a national landscape regarded by many as iconic, a landscape or set of landscapes, 'typically Australian in character', that allow the interplay of reflections on antiquity, heritage, identity and scientific knowledge, the guiding themes of this book. But unlike Hallett Cove or Lake Callabonna, the Willandra Lakes have entered a global canon of culturally significant landscapes. This has provided a larger stage for the performance of debates about truth, science, origin and ownership. Its global profile has helped reconfigure Lake Mungo as a place of pious pilgrimage and outstanding natural beauty and it has become emblematic of the repatriation debate in Australia.

The Willandra stories bring human beings decisively into the Pleistocene picture. With human presence comes human politics. The specific, intimate, personal stories of the Willandra Lakes' history and prehistory become incorporated into regional, community and academic debates, played out on national and international stages. Some archaeologists envisage a world free of Indigenous politics, and warn that there will come a day when people of Aboriginal or Native American ancestry, trained as archaeologists, wish to find out about their own destroyed prehistories. Those in favour of the repatriation and even the destruction of human remains have countered that the search for 'origins' in Australia is irrelevant and even counterproductive, as Aboriginal people know that they have always been here. Some archaeologists who are prepared to surrender all claim to historical burials are unwilling to relinquish their claim to the right to study 'prehistoric' remains. In turn, some Aboriginal people argue that all remains are ancestors, regardless of age, and there is no statute of limitations of the sacred. The past practices of naturalists, physical anthropologists and antiquarians cast long shadows over the controversy.

So far, sites like Hallett Cove and Lake Callabonna, with little or no obvious ancient material culture or human skeletal remains, stay relatively free of the angst and outrage that accompanies intellectual endeavour in the Willandra Lakes. However, they are far from blank slates on which Western scientists have inscribed geological knowledge. Incidents like the Broome trackways theft, and accounts like the *kadimakara* explanations of the Dieri people

Figure 41: Landscapes of the Willandra Lakes. (a) Lake Mulurulu lunette. *(Photograph K. Douglas).* (b) Joulni archaeological site, Lake Mungo. *(Photograph Michael Main)*

demonstrate that all Australian landscapes are layered with practical and spiritual significance for Aboriginal people.

The three threads of this essay track major themes of age, aridity and the watered inland; a social history of land management, interpretation and value; and the potentially intractable problems of archaeology and the disposition of material culture. These threads are united in the competing sacralisations of heritage, science, pastoralism and Aboriginal custodianship, and the coincidence of human occupation. In some sense each theme involves a battle for the control of meaning in the landscape. But meaning is not established in a vacuum by physical geography, painters, poets, farming methods, palaeontology, archaeology, local people, novelists or World Heritage guidelines. It is the cultural and intellectual overburden that distinguishes the Willandra Lakes Region World Heritage site, as it is viewed and imagined, from broad ovoid saltbush plains fringed by eroding dune deposits (see Figure 41). The power of deep time to add contour and colour to 'dead' landscapes is akin to the transformative effect of Technicolor on Dorothy's experience of the shift from Kansas to Oz, as we move from the eroding Mungo lunette, via the fourth dimension, to a fully treed and watered landscape: the Pleistocene land of lakes.[3]

Afterword

QUATERNARY SCIENCE, LANDSCAPE AND THE DEEP HUMAN PAST

Landscapes are the collection of physical and ecological features given meaning through a sense of beauty, and perspective of the past, and an understanding of the economic and cultural forces that shape the present ... the physical manifestation of people–land relationships that have evolved over many centuries[1]

A landscape is more than a set of physical, biotic and climatic attributes superimposed over bedrock. To leave it at three dimensions is to deny the richness of time and the human past. George Seddon wrote in 1996 that 'Human history is not complete without environmental history. It is not enough to detail the actions of the actors; the stage is equally important.'[2] Indeed, without the stage, we would have no awareness of the scale of the human action. Without recognition of the depth of ecological history, we could comprehend no deep human history.

Based on evidence from the ground beneath their feet, and on first-hand observation of the slow dissolution of the planet on human time scales, most European naturalists at the beginning of the nineteenth century accepted that the age of the planet was greater than the few thousand years prescribed by Biblical chronologies. By 1800, Georges Cuvier had proved the reality and apparent periodicity of the extinction of species in the fossil record, to his own satisfaction and eventually to the satisfaction of most of his peers, demonstrating that modern biota had a deep past. Further discoveries of extinct fossil remains during the century continued the challenge to Biblical literalism offered by the natural world, and propelled the study of vertebrate palaeontology into the public consciousness.[3]

In 1830 John Murray published the first volume of Charles Lyell's seminal *Principles of Geology*. Lyell's text set out to control the terms in which a discipline defined itself and in the process furnished a convincing model for the gradual shaping and reshaping of the planet over time, by causes now in operation. In the 1840s Louis Agassiz provided the bones of a new science, glaciology, which supplied a mechanism susceptible to appropriation by proponents of both gradual and 'catastrophic' change, to account for the shape and ecologies of post-Pleistocene landscapes in Europe and the Americas.[4] Before 1850, Charles Darwin had begun to flesh out the framework for a theory of the slow transmutation of species over time by natural selection, culminating in the first edition of *On the Origin of Species* in 1859. And the 1850s saw in the Western scientific canon the beginnings of the establishment of a great depth of human antiquity, perhaps epitomised from an Anglocentric perspective in the publication in 1865 of John Lubbock's *Prehistoric Times*. The world of lived experience suddenly had a deep ancestry.

With the relatively abrupt acceptance of a greater depth of time for human history, the counterfact of Seddon's observation emerged: Is it possible to study the history of the recent Earth without taking human actions into account? The stage is important, but without actors it is of limited interest to the audience. The theatre is a restrictive analogy: I do not wish to set up a dichotomy between humanity and nature, but merely to point out, with Seddon, that 'human' and 'environmental' histories are the same thing. The stage is an actor, not a passive backdrop. And not to put too anthropocentric a spin on it, late Quaternary history is at least in part the story of the interactions of human communities with their environments. In philosophical terms, acceptance of human prehistory opened an intellectual space for archaeology and physical anthropology to extend their compass into the deeper past. People became natural history objects.

In Australia, the age, affinities and origins of mainland and Tasmanian Aboriginal people were preoccupations of nineteenth-century geologists, comparative anatomists and physical anthropologists. They gleaned evidence for Aboriginal longevity or otherwise, and cultural stasis or adaptation, from the landscape, gathering the material remains of the past, including human burial remains, to support or dismantle theories, leaving a legacy of bitterness for their late twentieth-century intellectual successors and many modern Aboriginal people to negotiate.

The World Heritage significance of the Willandra Lakes region hinges upon this tension: the global and national importance and fascination of all 'prehistoric' human remains, versus their geographically and temporally specific cultural significance to Aboriginal Australians. In the words of Jim Bowler, the association of 'Ice Age environmental history with the story of human presence' at the lakes provides 'many aspects of modern social, even political relevance', as lines of demarcation are drawn in the sand dunes between scientific disciplines, institutions, pastoralists and Aboriginal people.[5] The Willandra Lakes share narrative resonances with the other landscapes as well as more concrete questions about protection and recognition of places worth saving. For example, Mungo's tale of decline from 'a 135sq km lake amid a temperate, forested Eden' where herds of monsters 'nibbled lush shoreline reeds' at the dawn of human presence evokes nineteenth-century imaginings of Lake Callabonna's 'huge trees', colossal beasts and 'dense tropical growth, exceeding in luxuriance the forests of the eastern slopes of the Andes', and the chimera of freshwater lakes in a dry country.[6]

This drying at Lake Mungo, and perhaps at Lake Callabonna, reflects glacial advance and retreat elsewhere in the world as the ice caps of the Pleistocene and the dust clouds of central Australia 'enveloped all nature in a shroud'. Hallett Cove provides a deeper ice age story. When Ralph Tate discovered the scarred pavements south of Adelaide in the mid-1870s, debate about the possibility and nature of glaciation in Australia (on which hinged contemporary understanding of the distribution of species) distracted from the issue of the antiquity of the Hallett Cove features. In the space of a decade, these features aged from 10 000 to 240 million years. There are no written records of explanations for the grooves on the rock pavement or the mysterious bus-sized boulders on the beach by the Kaurna people or their ancestors, but there is no doubt that the people who camped for millennia on the cliff tops incorporated this evidence of former worlds into their cosmologies. Quaternary change at the cove is represented by the striking erosional features of the Amphitheatre and memorialised in the tradition of Tjilbruke which accounts for the presence of freshwater springs along the coast. Ultimately, the significance of the site for Earth scientists alone was not enough to save it from urban expansion in the 1960s. Instead a combination of public and institutional interest and government lobbying at a particularly friendly moment in South Australia for heritage regulation led to its protection. Memories of visits to the cove by generations of South Australian schoolchildren and first-year geology students forced to endure excursions in all weather in the 70 years since its scientific discovery also contributed to the social value of the cove. The bureaucratic legacy of the debate ensures that the cove landscapes are doubly celebrated, for their geological significance and for their role as a test case for natural heritage protection in South Australia.

At Lake Callabonna, Aboriginal oral traditions about giant bones in the landscape resonate with palaeontological orthodoxy about change over time and cyclical climate shifts. Its protection, since 1901, is largely pragmatic, and reflects the limits of agriculture in South Australia as much as it does political concern with the status of fossils. The site, with its necropolis of Diprotodon and other megafaunal fossils, could be enormously important, were it not for the practical issues of isolation and accessibility, post-excavation preservation and reconstruction, and the associated costs.

Landscapes reverberate with cultural, political and intellectual meaning, and the deeper chronology that informs the history of human interaction with the 'natural' and the built environment. Furthermore, modern landscapes are as much social and political phenomena as they are physically constructed. As heritage specialists O'Riordan, Shadrake and Wood asserted in the quote which prefaces this concluding essay, echoing Seddon and Bowler, meaning in a landscape is established by 'a sense of beauty, and perspective on the past', as well as by other cultural and economic agents. Attitudes to place are formed by such perspectives. Modern landscapes are the 'physical manifestation of people–land relationships' evolving not over centuries, but over tens of millennia.

The importance of geological heritage lies in this conjunction of cultural and environmental histories with the Earth sciences. As well as a research and educational resource, a site of geological or archaeological heritage is a memorialisation of human and landscape interactions, both physical and intellectual, significant for the construction of identity and for community appreciation of the genetic influences on country. Perhaps understanding the land is the key to treating it better. Bowler and Seddon both stressed the importance of local geological knowledge to 'enhance the sense of place' to which we owe 'our sense of our individual and communal identities'. Bowler explained that 'as a geologist with a special interest in landscape evolution', he sees 'the constancy of change everywhere' when he travels. In 1996, Seddon noted that such an informed sense of belonging is fundamental to effective environmental planning and a key to 'feeling at home' in a landscape, while Bowler noted in 1995 that the discovery of the antiquity of Mungo Woman 'not only changed the dimension of Australian history but, in a special sense, it also changed the whole notion of being Australian', as 'realisation that this land nourished and was known by whole populations of people for 40 000 years instantly changed the nature of that landscape' and the ritual nature of the burial marked the place with 'the sense of spirit'.[7]

Putting down roots

The Antarctic explorer Charles Laseron wrote the bestselling *The Face of Australia* and *Ancient Australia* in the 1950s. Four decades later Simon Schama traced the 'ferocious enchantment' of national identity to the 'mystique of a particular landscape tradition'. Echoing Bowler and Seddon, Schama observed that geography, 'mapped, elaborated, and enriched as a homeland', in turn helps define the contours of national identity.[8] Laseron's introduction to *The Face of Australia* could have been Schama's exemplar:

> To those who have put down roots in the soil, what the pioneers found unattractive now binds us to the country. The unfamiliar has become the familiar; it has become part of us and we of it; it is the spirit of our national pride, the basis of our love of country.

Laseron wanted the highly urbanised Australian public to engage with Australia's varied and ancient landscapes. He sought a new denotation of landscape beauty, which could encompass the strangeness of Australia to non-Australian eyes. In 'so many lands', the post-glacial and alpine landscapes of North America and western Europe for example, or the rugged, tectonically excitable scenery of northern California, scenic beauty 'is an attribute of youth'. By contrast in Australia, he observed, 'it is expressive of great age': 'Its contours speak of vast antiquity, of long eras of stability, of freedom from the cataclysms that are still changing the shape of other countries.'

Like Laseron, the South Australian natural scientist and writer Hans Mincham appealed to and celebrated the material remains of the deep past as a national resource. He regarded it as crucial that his book about Australian Pleistocene mammals, *Vanished Giants of Australia*, reach a wide non-specialist audience, children as well as adults.[9] Published in 1967, *Vanished Giants* revisited many of the heroic foundations of Australian vertebrate palaeontology. Mincham commemorated as heroic not only the serendipitous discoveries and the planned expeditions to inhospitable locations, but the painstaking imaginative task of recreating an extinct animal from fragile, disarticulated bones. His book celebrated vertebrate palaeontology as a noble and national pedagogic endeavour, locating national pride in the material remains of a deep, non-human past.

Aboriginal oral traditions are also appropriated into the shared archive of nationally significant cultural productions. The myths, legends and folk tales of Mary Grant Bruce, Katherine Langloh Parker and others, incorporated into a transcultural corpus of tradition and literature that includes imported Anglo-Celtic cultural baggage, historical and semi-historical figures like Peter Lalor, Breaker Morant, Simpson and his donkey, Don Bradman, the Man from Snowy River, Ned Kelly and Captain Thunderbolt, and the Gum Nut Babies and Banksia Men of May Gibbs, provide non-Aboriginal Australians with stronger links with the landscape, further legitimising white Australian nationhood.[10] These 'traditions' are de-historicised and exclusionary. They are part of the celebration of a nationhood that is largely white, male, Christian and Anglophone.

The shared vocabulary of national or provincial identity is intimately tied to the physical or imagined/invented boundaries and spaces of a nation or a region, which includes the spaces of the past and the evocation of country as national heritage plays a further role in the nationing of the Australian landscape. Even a phrase like *the Australian landscape* is anomalous and anachronistic. Australian landscapes are neither homogeneous nor ineffably Australian through time. They have only been nationally 'Australian' for a little over 100 years. Their modern shapes are determined by up to 200 years of European-style cultivation with an evolving Australian twist. Before and alongside the British settlers, as much as 50 000 years of Aboriginal husbandry shaped different parts of the continent in different ways. Superimposed over and preceding the human influences are grander scales of climate and tectonically induced changes, and generations of biological agency. Indeed the continent itself may have taken on many different outlines in the last few hundred millennia, depending on the amount of water iced up at the poles, not to mention the intricacies of plate tectonics.

Earth scientists value a geological heritage site for its physical qualities or for the stories it tells about the development of their discipline. Geoheritage internationalises Australia at the same time as it provides a focus for national identity. So deep time and geological heritage offer a complementary transnational counterpoint to the national landscape. Geologically significant sites and their ongoing protection ensure the continuity and integrity of Earth science research in Australia. The landforms – as receptacles in which to locate geological knowledge – keep the history and the future of the discipline.

Protecting the deep past

The South Australian division of the Geological Society of Australia has played a central role in the preservation of geological heritage in that state since 1966 when the glacial pavements of Hallett Cove became the heritage movement's flagship.[11] From 1958–75 campaigners fought to prevent housing developers from building over coastal landforms and geological exposures with historical and educational significance. The society declared the glacial landforms a geological monument in the 1960s and the 50-hectare park which opened in 1976 was placed on the Register for the National Estate in 1981. The geological heritage subcommittee of the South Australian division has to date collated over 400 geological monuments in the state into the multi-volume Geological Monuments in South Australia. This project began as early as 1967 when subcommittee minutes mention Hallett Cove, Lake Callabonna and an Ediacara fossil reserve in the north of the state as geologically significant sites. The Hallett Cove campaigns of the 1960s and 1970s were initially driven by members of the GSA, the Geology Department of the University of Adelaide and the newly formed Nature Preservation Committee of the National Trust of South Australia. The perceived danger to Hallett Cove's geological integrity posed by the proposed suburban developments mobilised academic geologists, teachers and those concerned more generally with lack of public consultation and the state's natural heritage.

Lake Callabonna made an easier transition to protected landscape, declared a fossil reserve in 1901 (see Figure 42). But the protection offered by gazettal did not extend to practice. The fossil site was largely ignored, presumed 'lost' by local landholders and the South Australian Museum, for some 50 years, until the intervention of the international face of science, in the person of Ruben Stirton, then at Berkeley, who rediscovered the museum's original campsite and the necropolis of Diprotodon bones. Its relative obscurity, desolate

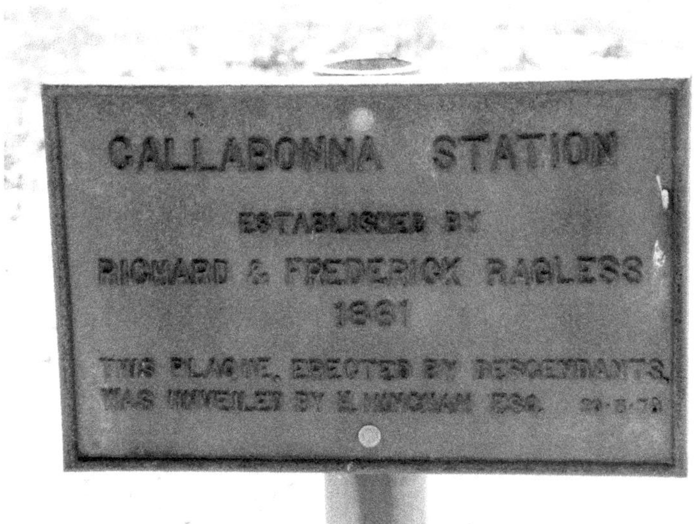

Figure 42: Callabonna homestead est. 1881. *(Commemorative sign at the ruins of Callabonna Homestead, erected by the Ragless family on the centenary of the establishment of the station by brothers Richard and Frederick Ragless in 1881 (Hans Mincham pers. comm.). Photograph John Magee)*

setting and isolation helped protect the site from the covetous eyes of developers (unlike at the conveniently-close-to-Adelaide Hallett Cove), poachers and fossil dealers, and indeed from the intrusive excavation techniques of professional palaeontologists.

Unappreciated by Earth scientists until the 1960s, Lake Mungo and the Willandra Lakes system entered the archaeological canon at the right moment for legislative protection. Gough Whitlam established the Hope Commission of Inquiry into the National Estate in 1972. By 1974, the Register for the National Estate was enshrined in the now superseded Australian Heritage Commission Act.[12] Australia was an early signatory to UNESCO's World Heritage Convention, which came into force late in 1975 with support from Malcolm Fraser's new conservative government, and the lakes had high profile advocates.[13] By 1979 Albert and Venda Barnes, the lessees at Mungo Station, had retired to Mildura, leaving land over which they had held custody for 45 years to allow gazettal of the new Mungo National Park, preliminary to inscription of the Willandra Lakes Region as Australia's fifth World Heritage property in 1981. The last of the three sites to receive formal protection for its ancient wonders is today a major tourist attraction, protected to the limit of conservation legislation as a National Park with World Heritage classification for its cultural and natural values, celebrated as an Australian Ararat, the site of the oldest human burial yet found in Australia.

How to recognise heritage

All this serves to suggest that deep time and the material remains of the past endow a landscape with value, mystery and grandeur, especially in a continent apparently lacking in scenery which corresponds to Western notions of the sublime or romantic. But a tract of land needs more than just a deep past to capture public, bureaucratic and heritage imaginations. Recognition – the perception of significance – precedes celebration and protection, which means that formal classification and appreciation are locked in an uncomfortable handshake. In turn, the identification and celebration of places might endanger the values for which they are recognised, by encouraging commercial exploitation in the form of tourism, development and, in the case of fossils or antiquities, removal.

Federalism has led to an Australian heritage regulatory framework which is multi-tiered, confusing and misunderstood. Federal and state governments can regulate for the protection of sites and features to a certain extent, but are hampered by the needs of resource sectors, the division of power between levels of government and the difficulty of systematically identifying places worth saving. In any case, it is not clear that it is the role of governments to protect every place with heritage significance, or that regulation is the best way to achieve protection. Scientists, journalists, politicians, settlers and Indigenous Australians have different heritage constructs. Furthermore, landscapes are larger than themselves. Meaning is not intrinsic, although it is shaped in part by the stories the world tells about itself. Value is established not only by geology, climate or biota, but the cargo of memory and representation that distinguishes them.

The declaration of a landscape as heritage draws on a powerful combination of nationalism and the sacred that renders the heritage values somehow immanent, already present, in the landscape. For example, Hallett Cove is a Mecca for geologists; Lake Callabonna is mysterious; Lake Mungo is scriptural. Nineteenth-century and European imaginaries provide context for their heritage recognition and depiction: a western European heritage weighs heavily on statements about 'Ancient Man', ruined walls, the notion of geological monuments and the Mosaic underpinnings or mediterranean imaginings of an inland sea. Reference to sites of world renown speak to uncertainty about claims founded solely on national significance. And there is an unpredictability to the canonisation of particular places and features, or their terms of reference (for example, Uluru is World Heritage listed for its outstanding geological significance, apparently because it is a large rock). Storytelling and human engagement may be the key that differentiates the heritage potential of particular sites from others that could have equivalent 'intrinsic' or scientific significance. At all these places, science helps constitute culture and *vice versa*. Scientific discourse helps shape the dimensions of public opinion, political will and modern conservation and interpretation priorities.

My aim has been to explore ideas about age in the landscape since the earliest public European acknowledgement of deep time in Australia, with the announcement of the Wellington Caves fossil discoveries in 1830. My major focus has been on the stories of discovery, interpretation, mobilisation and celebration of that deep past by geologists, palaeontologists, archaeologists, museum workers and their publics in the 170 years since then. This is an account of the development of the significance of the deep past in Australia through the emergence of particular deep-time places into the realm of heritage, and thereby into regional, national, and international frameworks of heritage recognition. People as well as ideas have passed through the places and narratives (Ralph Tate, Walter Howchin, Edgeworth David, J.W. Gregory, Gerard Krefft, Charles Sturt, Thomas Mitchell, Richard Owen, Charles Darwin, James Croll, Charles Lyell, Fred Ragless, Louis Agassiz, Edward Stirling, John Mulvaney, Henry Hurst, John Meldrum, Jackie Nolan, Norman Tindale, Maud McBriar, Isobel McBryde, Ruben Stirton, Julian Tenison Woods, Dick Tedford, George Seddon and Jim Bowler to name a few of the more prominent). Schama's claim in the monumental *Landscape and Memory* that the overwhelming rhetorical appeal of national identity is located in the 'mystique of a particular landscape tradition' is given weight by the attachment of Australians and popular media to deep-time objects and landscapes which become repositories for regional and national stories. Earth scientists and the popularisation of geology play an integral role in the creation and nurturing of Australian national or regional identities, through the use of the material remains of former worlds to transform country into a four-dimensional, time-travelling entity with a deep future as well as a deep past.

Endnotes

Introduction

1 Mulvaney and Kamminga, 1999; Moyal, 1986.

2 Howchin, 1901, 10; Mawson, 1936, lvi–lxv. Mawson's role in this debate is referred to in Seddon, 1996, 495. See also Hoffman and Schrag, 2000, 50–58; Veevers, 1990, 1–16; Mulvaney and Kamminga, 1999.

3 Griffiths, 1996; Reingold and Rothenberg, 1987; and others.

4 Hall, 1992, 234; Byrne, 1996, 82–107; Douglas, 2004; Douglas 2006; Robin 2007.

5 Rudwick, 1976.

6 Richard H. Tedford, Field journal, 'Australia expedition, 1953', Part One, 15 March to 29 June 1953, unpublished MS, private collection of Richard H. Tedford, 4 June 1953.

7 Miller, 1857, x–xi.

8 White, 1987, 2.

9 Seddon, 1996, 491, 490, 494.

10 Reproduced in Rudwick, 1995, 45. See page viii for his comment on 'crowding' a variety of organisms 'that would be unlikely to pose so obligingly together in real life'.

11 Plimer, 1997, 102–103. See Barthes, 1993, 100–102; Bowler, 1976; Gee, 2000, 5; Gould, 1996, 135–146; Gould, 1989; Kohn, 1996, 9–27.

12 Gee, 2000, 7–10; Seddon, 1996.

13 Sober, 1988, ix.

14 Darwin, 1861, 336.

15 Porter, 1977, 196–197; Young, 1985, 12–13, 127, 139; Moore, 1986, 322–326; Bacon, 1605, cited in Young, 1985, 12.

16 Benjamin, 1969, 257.

17 Fabian, 1983, 11–12.

18 Fabian 1983, 15–16.

19 Fabian 1983, 15–16.

20 Seddon, 1996, 489; McPhee, 1998, 29; Gee, 2000, 3–4, 26.

21 Seddon, 1996, 490–491.

22 McPhee, 1998, 115–137; Seddon, 1996, 494. See also Le Grand, 1988 for an account of continental drift and the plate tectonics revolution.

23 Douglas, 1998b, 16–17. Quotations are translated by Douglas from Roland Barthes, 'Le discours de l'histoire', *Information sur les sciences sociales* 6, 1967, 74.

24 Fabian, 1983, 15.

25 Griffiths, 2001, 23.

26 DEC, 2006.

27 Cf. Pyne, 1998.

28 Benjamin, 1969, 257.

29 Seddon, 1996, 493.

Part One

1 Douglas, 2006. This was communicated among other places in the pages of the *Transactions of the Royal Society of South Australia* in 1940 and 1942 and a number of Walter Howchin's publications during the early years of the twentieth century.
2 NPWS SA, no date, 1.
3 Giesecke (ed.), 1999, 7.

Chapter 1

1 Much of the geological information to follow is found in Cooper *et al.*, 1970 and Giesecke, 1999. Giesecke used the device of snapshots and Cooper *et al.*, scenes.
2 McPhee, 1998, 31.
3 For example, see discussion in Douglas, 2002, 68–83.
4 Bowler, 1976, 279–310; Bowler, 1986, 117–147.
5 Cooper, 1968b, 625; [Toteff], xv, 1986, 3, 66; Rendle, Major and McBriar, 2001, 11–12. Also Barry Cooper, pers. comm., 26 January 2000; Maud McBriar, pers. comm., 22 January 2000; Land Titles Office of South Australia, *Land Services History Book* vol. 35, Adelaide.
6 Douglas, 2004; Douglas, 2006.
7 Douglas, 2004; Douglas, 2006.
8 Tate, 1893, 31.
9 Dolling, 1981.
10 Evans *et al.*, 1986; Yelland, 1983.
11 Dolling, 1981; Evans *et al.*, 1986, 5.
12 Evans *et al.*, 1986, 5.

Chapter 2

1 Darwin to Lyell, [12 March?] 1847 (Cambridge University Library Darwin Papers DAR 146: 147).
2 Agassiz, 1886, 260–264; Agassiz, 1967; also cited in Bolles, 1999, 53–59, 87.
3 See Agassiz to Buckland in Agassiz, 1886, 289–290. Also quoted in Grayson, 1984, 11.
4 Buckland, 1842, 332–337, 345–348; see also letters and commentary in Agassiz, 1886, 290, 307–310.
5 Geikie to Darwin, 20 [April?] 1876 (DAR 165G, 1–40: 29).
6 Lyell to Wallace, 24 March 1869 (Wallace Papers, 113–128); Bolles, 1999, 49–51, 211, 217–223; Lyell, 1881, 245–280.
7 Darwin to Lyell, [8 September 1866], (DAR 146: 297).
8 Gray to Darwin, 27 August 1866 (DAR 165G: 154).
9 Darwin to Lyell, [7 February 1866] (DAR 146: 292).
10 Darwin to Lyell, [16 February 1866] (DAR 146: 293).
11 Croll, 1864, 121–137; Croll, 1866, 301–305; Croll, 1867, 426–445; Croll, 1869, 201–208. For biographical detail see Gribbin and Gribbin, 2001, 31.
12 Croll, 1867, 430–445.
13 Croll to Darwin, 2 December 1868 (DAR 50 series 5).
14 Croll to Darwin, 2 December 1868 (DAR 50 series 5); Croll, 1866, 305.
15 Scoular, 1879; Tate, [1894?–1901], 34; Wallace Papers and Darwin Correspondence between Lyell, Wallace, Croll and Darwin, 1868–1869.
16 Darwin to Croll, 24 November 1868 (DAR 143).
17 Gribbin and Gribbin, 2001, 40.

18 Gribbin and Gribbin, 2001, 50–57.
19 Darwin to Croll, 24 November 1868 (DAR 143).
20 Veevers, 1990, 1–16.
21 Hoffman and Schrag, 2000, 50–58.
22 Dr Brett Marmo, email, 17 March 2004.
23 The bulk of the information above on glacial super-cycles and modern orthodoxy on ice ages has been derived from Gribbin and Gribbin, 2001; Hoffmann and Schrag, 2000, 50–58; John, ed., 1979; Maasch, 2003; Veevers, 1990, 1–15; Stephen J. Pyne presented an Antarctic perspective in Pyne, 1986, 289–312.
24 Scoular, 1879, 66–69; Tate, [1894?–1901].
25 For example, see Selwyn, 1859, 4; Fenner, 1937, v; Tate, 1879, lxiv.
26 Clarke, 1853, 37–38; Clarke, 1860, 86, 94, 105, 117, 167, 221–234; Clarke to Darwin, August 1861 (DAR 161.2: 171). Aplin to Clarke, 9 September 1863: 'I am only a *Govt. prospector* & not Govt. Geologist at all', quoted in Moyal, 1986, 124–127; Scott, 1977, 72, 75, 89–91; Griffiths, 1884, 7–10.
27 Haast, 1867, 278.
28 Woods, 1867, 43–47.
29 Ellery, 1868, 3; Haast, 1867, 273–278; Woods, 1867, 44, 47.
30 Woods, 1876, 78.
31 Player, 1992, 108.
32 Lane, 1962, 15–20.
33 Woods, 1867, 46, 47.
34 Woods, 1883, 382. See also Woods, 1867, 44, 45, 46; Ellery, 1868, 2–3; Clarke, 1860, 167; Clarke to Darwin, August 1861 (DAR 161.2: 171); Tate, [1894?–1901], 33.
35 Woods, 1862, 20; Woods, 1883, 381–382; Tate, 1879, lxv, lxvi.
36 Mawson, 1936, lxiii; Scott, 1977, 72; Tate, 1879, lxiv.
37 Selwyn, 1859, 4.
38 Selwyn and Ulrich, 1866; my italics.
39 Taylor, 1873–1875, interleaved with Selwyn and Ulrich, 1866, facing page 9.
40 For example, Tate, 1879, lxiv; Tate, 1893, 31.
41 Tate, 1879, xl.
42 Tate, 1879, lxv–lxvi. See also Scoular, 1879, 60–70; Scoular, 1885, 36–48.
43 Anon., 1884, 116–117; Player, 1990, 108.
44 Cooper, 1992. Between 1878 – aged 73 – and his death in 1887, Scoular published eight papers in the Royal Society of South Australia's *Transactions*.
45 Scoular, 1885; Tate, 1885.
46 Bowler, 1976, 279–310.
47 Hutton, 1885, 334–335, 339; Lendenfeld, 1885, 45–46, 52.
48 Scott, 1977, 91; Lendenfeld, 1885, 53.
49 Hutton, 1885, 341.
50 Tate predicted that on viewing the Hallett Cove features, glaciologically informed scientists could not fail to be convinced of their origins as reported in the *Observer* and Tate, 1893, 31. General acceptance of their glacigene nature is acknowledged in Howchin, 1895, 61. See also David, 1895, 58: 'To-day ... if any geologist were to question the glacial origin of the pavement at Hallett Cove, he would be looked upon by his fellows as a dangerous lunatic. Thus has the whirligig of time brought in revenges to the old prophets.'
51 Howchin, 1895, 62; Tate *et al.*, 1895, 315–320; Tate, 1895, 276.

52 Joyce, pers. comm. 23 February 2000. Much of the work of the Australasian Glacial Evidence Committee was taken over by the Cainozoic and Quaternary Climate Committee during the 1920s.

53 Howchin, 1895, 68–69.

54 Howchin, 1895, 69.

55 Anon., 6 April 1895, 58; Howchin, 1895, 67; David, 1895, 82, 84–88, 92, 94–95; Tate *et al.*, 1895; Bourman and Alley, 1990, 70–81. Bourman and Alley (1995, 67) stated that 'No marine facies has been detected at Hallett Cove', in a centennial confirmation of Howchin's assertion.

56 David, 1895, 58.

57 Revisions to Tate, [1894?–1901], vol. 1; Howchin, 1907, 264–272. For the long-running Howchin–Basedow dispute, see W. Howchin, 'Mr H. Basedow and the glacial beds of Cambrian age in South Australia', unpublished MS, [1911?], MSS 9, series 1.1, W. Howchin, 'The Basedow–Howchin controversy 1909–1911', in Sir D. Mawson, 'On South Australian geology', Barr Smith Special Collections, University of Adelaide; Illife and Basedow, 1905, 334; Iliffe and Basedow, 1908, 260.

58 Wheaton, 1937, 1.

59 For a statement on the importance of Howchin's work on the Sturtian tillite, see T.W.E. David (Sydney) to Professor J.W. Gregory, President, Geological Society (Glasgow, Scotland), 20 July 1928, Edgeworth David Papers, University of Sydney Archives, Series 16. My final sentence borrows from McPhee 1998, 104.

Chapter 3

1 South Australian *Planning and Development Act* 1966–67, sections 36–38. For detail see also unpublished paper on environmental law at Hallett Cove by Hamilton (1972, 2–5) and the Kadima and Silesia development companies' 'Hallett Cove development application' (T.S. Martin and Associates, 1971), in the Hallett Cove Papers.

2 Major Open Space legislation cited in Anon., 20 September 1971. See also Hallett Cove Papers; [Toteff], 1986, 14; Warburton and McBriar, 1975, 2–3. Quotes from [Toteff], 1986, xv; Bonython *et al.*, 1975, np; Ellis and Warnes, 1977, 5; Warburton and McBriar, 1975, 5.

3 Rudman *et al.*, 1975; NPWS, 'Hallett Cove Conservation Park'.

4 Bonyhady, 1996, 145; Robin, 1994, 1–24; Robin, 1998a, 199–202.

5 Caldicott and Geering, 1975, 113.

6 Bonython *et al.*, 1975, np.

7 Clarke, 1975, 1.

8 Clarke, 1975, 6.

9 Clarke, 1975, 13.

10 See 'cautionary note' from C.W. Bonython, MS of minutes of the second meeting of the National Trust's Subcommittee on Hallett Cove, 19 December [1966]: 'Apologies from Mr Bonython & cautionary note advising not asking too much in case get nothing' (Hallett Cove Papers).

11 T.S. Martin and Associates, 1971, 9–10.

12 Quotes from Caldicott and Geering, 1975, 113; T.S. Martin and Associates, 1971, 3; Anon., 15 September 1971, 1.

13 Robin, 1994; Robin, 1998a, 197–199; Robin, 1998b; Bernie Joyce, pers. comm., 23 February 2000; McBriar, pers. comm., 28 January 2000.

14 For example Hasenhohr, 1991/2 and Giesecke, 1999; Ice Age Trail Alliance, online <www.iceagetrail.org>.

15 For quotations see SASTA, 1974; Caldicott and Geering, 1974; McDonald *et al.*, 1965, np. For development of these ideas see Douglas, 2006.

16 Robin, 1998b, 90–112.

17 Robin, 1998b. For development of these ideas and full references see Douglas, 2006.

18 [Cooper], 1977.

19 Joyce, pers. comm., 23 February 2000; McBriar, pers. comm., 28 January 2000.

20 Geological Monuments Subcommittee annual reports, 1968, 1971 (Hallett Cove Papers).

21 For example, see Henk Godthelp quote in Hoy, 2001.

22 Geological Monuments Subcommittee annual reports, 1969, 1970 (Hallett Cove Papers); McBriar, pers. comm., 28 January 2000.

Hallett Cove and geological knowledge

1 David, 1895, 58.

Part Two

1 Reports on the weather and fossils from the *Register* and the *Observer,* 29 October to 5 November 1892.

2 *Advertiser,* 6 September 1893.

3 See Pledge, 1994; Rich and van Tets, 1985; and Richard H. Tedford, pers. comm., 20 July 2000, for examples.

4 Morton, 1996, xii, 16–18; Moyal, 1986, 94, 162; *Observer* and *Register* 5 November 1892.

Chapter 4

1 South Australian Government, 1901, 1117.

2 Stirling, 1907; Pledge, 1994, 75.

3 Eyre, 1845; Black, 1962, 43–51; E.C. Stirling, Minutes of the South Australian Museum [SAM] Board, 2 February 1894. Manning, 1990, 61 recorded that Mulligan is 'a corruption of the Aboriginal *mullachan* – "springs of water running". The name "Callabonna" was suggested by Dr Stirling in 1894 who failed to record its meaning.'

4 Basalla, 1967, 1–16. Roy MacLeod provided a contemporary critique of Basalla as applied to Australian contexts in his important 1987 paper, 'On visiting the "moving metropolis": Reflections on the architecture of imperial science', pp. 1–16. I have discussed Basalla's and others' models as applied to the Lake Callabonna locality in Douglas, 2001, 60–61.

5 Rod Wells, pers. comm., 3 February 2000; Richard H. Tedford, pers. comm., 20 July 2000. The quote is from Stirling, 1894, 184.

6 Tedford, pers. comm., 20 July 2000; Tedford, 1985, 39–59; See also the UCMP Collection – correspondence, field journals, sketches and miscellaneous clippings, MSS and TSS from the University of California Museum of Paleontology, UC Berkeley, especially the field journals of R.A. Stirton, R.H. Tedford, Alden Miller, M.O. Woodburne, W. Clemens. See articles like Frank Kennedy, 'Search next year for the daddy of them all' [*Advertiser?* (Adelaide)], 23 September 1970, unprovenanced newspaper clipping, Callabonna papers.

7 Richard H. Tedford, Field journal, 'Australia Expedition, 1953', Part One extract, March 14 to June 28, unpublished MS, 27 March 1953, 89 (private collection of Richard H. Tedford).

8 The quote is from Tedford, 1973, 352–353. Also see Martin Glaessner to R.A. Stirton, no date (UCMP Collection); Paul Lawson, pers. comm., 19 March 2001; Stirton, 1954, 6–8; R.A. Stirton, Field journal, 'February 12, 1953 to August 23, 1953, pp. 1 to 263', unpublished TS (UCMP Collection), 1 June 1953; Tedford, 1985, 39–59.

9 Tedford, pers. comm., 20 July 2000.

10 Fletcher, 1996, 150–152. Fletcher's memoir, written in 1996, 50 years after the event – records the year of the expedition as 1948, but as this memo demonstrates, he travelled to Lake Callabonna in

1945: 'On Friday next, 1st June, Mr H.O. Fletcher, Palaeontologist of this Museum, is proceeding to the Lake Callabonna district to do some field collecting for the Museum collections' (Walkom, 1945).

11 Sprigg, 1991, 11; Rod Wells, email, 5 February 2004. See also Smith, 2001, 8; and White, 2001, 26: The first Cretaceous mammal recovered from Australia was identified by Dr Alex Ritchie of the Australian Museum from an opalised jawbone found at Lightning Ridge in the mid 1980s. *Steropodon galmani* is a cat-sized monotreme which inhabited central Australia at least 100 million years ago. In 1992 another opalised monotreme jawbone was identified from the same area and named *Kollikodon ritchiei.*

12 Wells and Callen, 1986, 98–100.

13 Wells and Callen, 1986, 6, 99.

14 Wells and Callen, 1986, 99–100.

15 Macumber, 1969, 39–57.

16 Wells and Callen, 1986, 101.

17 Bowler, 2002; Bowler, 1976, 279–310; Wells and Callen, 1986.

18 Wells and Tedford, 1995; Bureau of Meteorology.

19 Wells and Tedford, 1995; Bureau of Meteorology.

Chapter 5

1 Rudwick, 2005; McPhee, 1998; Bynum, Browne and Porter, 1983, 422: entry for 'time (geological)'; Hutton quoted in McPhee, 1998, 140.

2 For example, Lyell, 1830; Buckland, 1823; Agassiz, 1840. See Rudwick, 1997, for a recent analysis and translation of some of Cuvier's papers.

3 Grayson, 1983, Chapters 4–6.

4 For an interesting account of monsters, fossils and nationalism in North America see Semonin, 2000.

5 Translated in Rudwick, 1997, 18–24.

6 Bednall, 1878, 77; Hoffman, 1989, 21–35.

7 Hoffman, 1989, 22; Rudwick, 1976, 156; Winchester, 2001.

8 For development of these ideas see Douglas, 1998a; and Douglas, 2002, 68–83. See also Ryan, 1996, 121.

9 Rich, 1985b, 20–21; Rich, 1985a, 191–194.

10 Mitchell Papers, C9A 295 – 3, 174; Field note and sketch book 1828–1830.

11 For example, Semonin, 2000.

12 For discussion of the use of Antipodean remains and informants in the formulation of metropolitan science see: Butcher, 1989, 1–15; Dugan, 1980, 265–272; Moyal, 1975, 41–56; Player, 1992, 107–109; Stafford, 1989; Vickers-Rich and Archbold, 1991, 5–10; MacLeod, 1987.

13 Mitchell Papers, draft of a letter to the Geological Society dated 2 September 1831.

14 Mitchell Papers, volume 8, 2 September 1831.

15 Quote from Mitchell to Owen, 28 January 1843 [copy], Owen Papers, Mitchell. For a discussion of Darwin's possible use of Australian fossil vertebrate material in the formulation of the law of succession of types, see Dugan, 1980, 265–272.

16 Ferdinand von Mueller (Melbourne) to Richard Owen (Royal College of Surgeons, London), 1862?, Richard Owen Papers, von Mueller, Volume 19, August 1861 to December 1887, Natural History Museum, London (henceforward Owen Papers, Mueller).

17 Mitchell to Owen, 28 January 1843 [copy], Owen Papers, Mitchell.

18 A copy of the original letter is in the Mitchell Papers, volume 8, State Library of NSW, Sydney. An undated draft also appears in T.L. Mitchell, Miscellaneous Papers, volume 6, State Library of NSW, Sydney.

19 Richard Owen to Roderick Murchison, 8 February 1845, LDGSL 838/07/6 (Murchison Papers).

20 Mitchell, 1839, volume 2, 368–369 and plates. See figures 1 and 2, Pl. 49.

21 Rich, 1985a, 188–194.

22 Letter from R.C. Gunn (Launceston) to E.C. Hobson (South Yarra), 3 July 1847, Box 865/2B, Hobson Papers, MS8457, La Trobe Collection, State Library of Victoria: 'I have *fabricated* an extract from an imaginary letter of yours so as to introduce the two species of Wombat'.

23 For example, see correspondence between E.C. Hobson (London, Van Diemen's Land and Port Phillip District), and Richard Owen (Royal College of Surgeons, London), 7 December 1836 to 25 May 1845, Box 865/1C, Hobson Papers, MS8457, La Trobe Collection, State Library of Victoria.

24 Orchiston *et al.*, 1977, 115. For references by Stirling and Zietz to the discovery and naming of *Genyornis*, see Stirling and Zietz, 1896a, 171–190; Stirling and Zietz, 1896b, 191–211; Stirling and Zietz, 1896c, 430, 593.

25 Owen, 1845. The work comprises 32 quarto pages discussing such Australian fossil genera as had been collected up to 1845.

26 Minutes of the Board of Curators, Volume C, 1844–1854, 'Meeting of the Museum Committee on the 16th of July 1846', Archive of the Royal College of Surgeons, London, 128–30.

27 Clarke, 1847.

28 Krefft to Owen, 11 June 1870 [copy], Owen Papers, Krefft. See also Krefft, 1873, 145.

29 Gerard Krefft, Correspondence, 1861–1878, Albert Gunther papers, unpublished TS, Australian Museum ref. AMS48, Sydney.

30 For example, see Krefft to W.B. Clarke, 19 May 1870, W.B. Clarke Papers, Mitchell Library, Sydney. See also Flower, 1868, 307–319, for his claim that Thylacoleo was herbivorous and subsisted on soft, ripe fruit.

Chapter 6

1 Anon., 31 July 1893, 6.

2 Hurst, 1893.

3 For hints of disputes, see Hurst, 1893: 'Mr Ragless who previously promised to lend me a horse, dray and water tank whenever I require them was unable to do so, but offered to hire his bullock dray at £3 per day if I could get a tank elsewhere – not considering this offer sufficiently moderate, I retained 3 camels pending your reply to my request for a further loan of them. Should this request not be granted I shall have to accept the terms of Mr Ragless.' See also Stirling, confidential letter to the Board of Governors, 28 August 1893, Callabonna Papers, SAM. The director claimed that Hurst 'troubled' the Ragless brothers 'by too frequent visits and requests for rations by the party'.

4 For example, Mincham, 1967, 41–43; Pledge, 1994, 68; Vickers-Rich and Archbold, 1991, 28.

5 For Hurst's correspondence see Hurst to Stirling, 11 March, 16 March, 6 May, 20 June 1893, Callabonna Papers. For quote see Anon., 31 July 1893, 6.

6 See Hurst to Stirling, 16 February, 11 March, 16 March, 20 March, 5 April, 6 May, 20 June; see also Hurst, 1893 and Hurst to Stirling, 11 July 1893, en route to Adelaide having left his brother in charge, expressing his concerns about injury to the specimens, which he was forced to transport by horse and buggy, as the camels were not available, Callabonna Papers.

7 Stirling, confidential letter to the Board of Governors.

8 Stirling and Zietz, 1899b; Pledge, 1994, 73.

9 Hurst to Stirling, 19 October 1918, with addendum from E.R. Waite (South Australian Museum, Adelaide) letter 26481, 018/3255 (private collection of R.H. Tedford).

10 Hurst to Stirling, 19 October 1918, with Waite addendum (private collection of R.H. Tedford).
11 These ideas are developed in Douglas, 2006.
12 Anon., 12 September 1894, 5.
13 Anon., 24 May 1893, 4–5.
14 For example, see reports Anon., 23 May 1893, 5; *Observer*, 27 May 1893, 43.
15 Ward, 1893, 4.
16 Owen, 1877.
17 For example, see Curator and Secretary [Gerard Krefft], Australian Museum, to the Colonial Secretary [John Robertson], Sydney, 10 May 1869, No. 2584, Shelf 173HG 13.3, Australian Museum Library collection; Krefft to George Bennett, 23 February 1870 [copy], Owen Papers, Krefft; Krefft to Richard Owen, 11 June 1870 [copy], Owen Papers, Krefft.
18 Stirling, 1893, 784.
19 E.C. Stirling (Adelaide Museum) to Professor E. Ray Lankester (British Museum, London), 14 February 1899, Callabonna Papers, SAM, Adelaide.
20 Gregory, 1906. Also correspondence between J.W. Gregory, A.H.C. Zietz and the South Australian museum board from September to December 1901, Callabonna Papers.
21 J.W. Gregory (University of Melbourne) to A.H.C. Zietz (Adelaide Museum), 14 September 1901, Callabonna Papers.
22 Gregory to Zietz, 2 October 1901, Callabonna Papers.
23 J.W. Gregory (University of Melbourne) to 'the Director of the S. Australian Museum' [E.C. Stirling], 1 December 1901, Callabonna Papers.
24 Brown, 1893, 6.
25 A.H.C. Zietz (Adelaide Museum) to E.C. Stirling (England), May 1901; Zietz to Stirling, 15 March 1901, Box 2, Folder 1, AA 309, SAM, Adelaide. Assorted correspondence on museum exchanges and reports of the museum board in the *South Australian Register* in Box 3, Folder 9 and Box 2, Folder 1, AA 309, SAM, Adelaide; Callabonna Papers; E.S. Booth Papers, State Library of South Australia, Adelaide.
26 Memorandum by A.C. Gregory on the provincial division of the northern portion of the Australian continent, from *Votes and Proceedings of the Queensland Parliament*, 1861. Quoted in Cumpston, 1972, 117; Ragless, no date.
27 Stirling and Zietz, 1913a, 111–126; Stirling and Zietz, 1913b, 127–178; Anon., 1919, 1–2.
28 Waite, 1917, 407; Hurst to Stirling, 19 October 1918, with addendum from E.R. Waite (South Australian Museum, Adelaide), (private collection of R.H. Tedford).

Chapter 7

1 R.A. Stirton to Whitney H. Shepardson, the Dominions and Colonial Fund of the Carnegie Corporation, 20 December, 1948 (UCMP Collection).
2 R.A. Stirton to the Max C. Fleischmann Foundation of Nevada, 6 January 1954 (UCMP Collection).
3 Stirton, 1954, 4–5; Stirton, Field journal 1953, 15 March–2 April 1953; R.A. Stirton to Charles L. Camp, Museum of Palaeontology, Berkeley, 19 August 1953 (UCMP Collection); Tedford, Field journal 1953, Part One, 15 March to 2 April 1953, 57–101.
4 Stirton, 1954, 4.
5 Quote from R.A. Stirton to the Max C. Fleischmann Foundation, 6 January 1954 (UCMP Collection). See also Stirton, 1954, 6, and R.A. Stirton to Charles L. Camp, 19 August 1953 (UCMP Collection).
6 Tedford, 1985, 40.

7 Stirton, Field journal 1953, 31 May 1953.
8 Stirton, Field journal 1953, 1 June 1953.
9 Stirton, Field journal 1953, 1 June 1953.
10 Tedford, Field journal 1953, Part One, 15 March to 29 June 1953, entry for May 31. 'VP' is vertebrate palaeontology.
11 Stirling and Zietz, 1900, 41–80; Stirling and Zietz, 1905, 81–110; Stirling and Zietz, 1913a, 111–126; Stirling and Zietz, 1913b, 127–178 (*Memoirs of the Royal Society of South Australia,* Parts 2–5, 1900, 1905, 1913).
12 R.A. Stirton, Field journal 1958, South Australia, 14 June 1958.
13 Stirling, 1894, 209.
14 Stirton, Field journal 1958, South Australia, 25 July 1958.
15 Stirton, Field journal 1958, South Australia, 13–14 August 1958.
16 White, 1973.
17 Stirton, 1954, 13; Stirton to Hale, 28 June 1955 (UCMP Collection).
18 Stirton, 1954, 11, 4. Letters from 1948 demonstrate a longstanding interest in Australia's fossil marsupial record, and its links to his Columbia project. See Stirton to Shepardson (Carnegie Corporation), 30 December 1948; and Annie M. Alexander (a benefactor of the UC Museum of Paleontology and keen amateur palaeontologist) to Stirton, 21 December 1948, on 'the Australian project', (UCMP Collection).
19 Stirton, Notebook 1954, [?] June 1954. [p. 42]
20 Paul Lawson, pers. comm., 19 March 2001.
21 Stirton, 'Notes on Australian fossils in various museum collections', 1957, item 5: Notes on a conference at the National Museum of the United States of America (Smithsonian Institution), Washington DC, 3 November 1957 (UCMP Collection).
22 Stirton, 1954, 11.
23 Tedford, pers. comm., 20 July 2000; Tedford, 1973, 353; Wells and Tedford, 1995, 3–5.
24 Tedford, pers. comm., 20 July 2000; Neville Pledge, pers. comm., 27 January 2000. See for example, *Adelaide Advertiser* 20 June, 2 July, 18 July, 31 August, 19 September, 21 September, 23 September 1970; *Sunday Mail* 5 September 1970.
25 Stirling, 1907; Long and Mackness, 1994, 95.

Skeletons in the cabinet and the lost world of Lake Collabonna

1 Rod Wells (email, 5 February 2004) observed that he has always estimated their age at a relatively young 70–75 000 years before the present. This still places them well outside the standard 45 000 year estimate for the oldest human remains found in Australia to date.
2 Stirton, Field journal 1958, 28 July 1958.
3 Hoy, 2001.

Part Three

1 For a reflection on parallels between nineteenth-century biblical and late twentieth-century secular scientific visions of the watered inland of New South Wales, see Douglas, 2002, 68–83.
2 Jim Bowler, pers. comm., 26 April 1999.
3 For example see the debate played out in the pages of the *Journal of Human Evolution* in 1999–2000: Thorne *et al.*, 1999, 591–612; Bowler and Magee, 2000, 719–726; Gillespie and Roberts, 2000, 727–732; Grün *et al.*, 2000, 733–741; DEC 2006.
4 Bowler, 1998, 154.

Chapter 8

1 Pardoe, 1995, 696; Brown, 1989, 127–146; Dawson, 1992.
2 Bowler and Magee, 1978, 8–10.
3 Douglas, 1996, 26–27, 35; Bowler, 1982, 36–37, 41.
4 Bowler, 1982, 37, 40; Brown and Stephenson, 1991; Brown, 1989, 127–146.
5 Bowler, 1976, 282–283; Bowler, 1982, 36, 38–39; Bowler and Jones, 1979, 679–685; Bowler, 1986, 117–147.
6 Bowler, 1986, 117–147.
7 Douglas and Ferguson, 1988; Douglas, 1996.
8 Bowler *et al.*, 1976, 359–394; Street and Grove, 1979, 99–103; Bowler, 1976; Bowler and Jones, 1979, 679–685; Bowler, 1990, 108.
9 Mitchell, 1839, vol. 2, 146–147, 154.
10 J.M. Bowler, pers. comm., 26 April 1999; DEC 2006.
11 Quotes from Smith, 1997, 88–89 and Our Art Critic, 1945, 9.
12 Newman, 1944a, 5.
13 Bonyhady, 1997, 12.
14 Newman, 1944a, 5.
15 Caption to photo, Newman and Leonard, 1944, 6.
16 Guy, 2003a, 147; see also Guy, 2003b; Levell and Wee, 2009, 48.
17 White, 1997, 91.
18 Bowler and Magee, 2000, 719–726; Jim Bowler, pers. comm., 25 July 2002; Douglas, 1996, see Figure 17 for example.
19 Newman, 1944a, 5.
20 Johnston and Clark, 1998, 105.
21 Mitchell, 1839, vol. 1, 373.
22 Hills, 1940, 15–21; Stephens and Crocker, 1946, 302–312.
23 Newman 1944b, 5.
24 Taylor, 1944, 8.
25 Hal Missingham quoted in Smith, 1997, 88–89.
26 Bonyhady, 1997, 10.
27 Newman, 1944b, 5.
28 Bowler, 1998, 120.

Chapter 9

1 Howitt, 1904, 51. See also Cameron, 1885, 346–347; Horton, 1994, 96; Tindale, 1974.
2 For example, Allen, 1972; Allen, 1974, 309–322; Mitchell, 1839, vol. 1, 253, 262, 264–275, 321, vol. 2, 71, 101–104, 113, 149.
3 Mulvaney and Kamminga, 1999, 199.
4 Bowler *et al.*, 1970, 39–60; Jones, 1973, 278–281; Allen *et al.*, 2008; Allen and Holdaway, 2009.
5 Colley, 2002, 13–14; Allen, 1998, 207–220; Allen *et al.*, 2008, 11–24; Allen and Holdaway, 2009; Gregory 1906.
6 Jose, 1997.
7 James Gleeson, 'Russell Drysdale', *Art Gallery of New South Wales Quarterly,* Sydney, October 1960, 42, quoted in Smith, 1997, 22; Anon., 16 August 2003, 6; Guy, 2003a, 6; Levell and Wee, 2009, 46.
8 Grayson, 1983, 1.

9 Marsden, 1984, 40; Trautmann, 1991; Lyell, 1859, 93–95. See also Lyell, 1970, 106, 185.
10 For example, see Marsden, 1984; Daniel, 1981.
11 Grayson, 1983, 14; Marsden, 1984, 45; McBryde, 1996, 66–68.
12 Marsden, 1984, 46–47.
13 Douglas, 1998b; Douglas and Ballard, 2008.
14 Dampier, 1906a, 453: Western Australia, January 1688.
15 Dampier, 1906b, 440: Western Australia, August 1699.
16 Tylor, 1894.
17 R.C. Gunn (Van Diemen's Land) to E.C. Hobson (South Yarra), 3 July 1847, Box 865/2B, Hobson Papers, MS8457, La Trobe Collection, State Library of Victoria, Melbourne.
18 Howitt, 1904, 33; David, 1925; E.D. Gill, Notebook 14, in Gill, Papers, PA884, Notebooks 1–58, La Trobe Collection, State Library of Victoria, Melbourne, 1963–1964, 46–149.
19 Mulvaney and Kamminga, 1999, 40–42; Colley, 2002, 2; Griffiths, 1996.
20 Bowler *et al.*, 1972, 48–50; Bowler 2005; Mulvaney 2005; Tom Uren quoted in Hughes 1996.
21 Thorne, 1971, 85–89; Thorne and Macumber, 1972, 316–319; Thorne, 1977; Stone, 2004, 18–20; Brown, 1981, 156–167; Pardoe, 1995, 703; Brown, 2000, 743–749; Grose, 2000, 3. See Mulvaney and Kamminga, 1999, 156–170 for an analysis and summary of the debate c. 1999.
22 Bowler, 1998, 120–155; Bowler and Price, 1998, 156–168.
23 Thorne *et al.*, 1999, 591–612. See also Anon., 20 May 1999; Grose, 1999; MacDonald, 1999; Woodford, 1999.
24 Grose, 1999; Bowler and Magee, 2000, 719–726; Gillespie and Roberts, 2000, 727–732; Grün *et al.*, 2000, 733–741.
25 Bowler, 2000, 11.
26 J.M. Bowler, pers. comm., 22 July 1999, Melbourne. See also Colley, 2002; Harris, 2001, 17, 20; Douglas, 1996.
27 For perceptions of the lakes' importance in this respect see, for instance, Tim Lloyd, 2003, 35 writing about Tim Flannery, then Director of the South Australian Museum, in the *Advertiser*. For a polarised introduction to the megafaunal debates, see Flannery, 1994; Horton, 2000; Ward, 1997.
28 Chippindale, 1985.
29 For more comprehensive recent discussion, see Colley, 2002; Fforde *et al.*, 2002; Griffiths, 1996, 94–100; McBryde, 1996, 79–83; Mulvaney and Kamminga, 1999, chapter 1, especially 8–10. See also the findings of the Human Remains Working Group, established in May 2001 by Alan Howarth, then Minister for the Arts in Britain, chaired by Professor Norman Palmer: *Report of the Working Group on Human Remains in Museum Collections* (Palmer Report).
30 Fforde *et al.*, 2002, 2–3.
31 Renfrew and Bahn, 1991, 463; For more detail on 'indigenisation' and appropriation see Byrne 1996, 82–107, and Carter, 1995, 3–15.
32 McBryde, 1996, 74. See Wallis *et al.*, 2008 for an example of collaborative work in South Australia.
33 McBryde, 1996, 67; Renfrew and Bahn, 1991, 483. See for example an article by Professor Mulvaney discussing the 'handback' of Kow Swamp remains ('Bones of contention', *Bulletin,* 9 October, 1990, 104–105). The *Bulletin* article anticipated a longer piece that appeared in *Antiquity* (Mulvaney, 1991, 12–21).
34 Jopson, 1999; McBryde, 1996, 66.
35 See for example Johnson, 1996; Wilford, 1999; Egan, 2000. The perspective of many archaeologists, personified by the Smithsonian's Doug Owsley, is reflected in Jeff Benedict's account of the Kennewick Man controversy in Benedict, 2003.
36 Benedict, 2003; Renfrew and Bahn, 1991, 9. My italics.

37 Dillon, 2004, 18.

38 Howitt and Siebert, 1902, 525–532; Gregory, 1906, 3–9; Fiffer, 2000; Larson and Donnan, 2002; Reynolds, 1999; Nedin, 1996. An account of the trackways incident, and the long hunt to recover the trackways, appears in Long, 2002. The theft was reported in the *Australian* on 16 October 1996.

39 Jose, 1997, 354.

40 Byrne, 1996.

41 Bowler, 1998, 120.

42 For opposing views see Johnson, 1996, and Jopson, 1999, 5. See also Egan, 2000; McBryde, 1985; McBryde, 1996, 63–84; Mulvaney, 1990, 104–105; Wilford, 1999.

43 For example, see Johnson, 1996; Appleton, 2002; Levitt, 2001.

44 Mulvaney, 1991a, 12–13.

45 Mulvaney, 1991a, 19.

46 Mulvaney, 1991a, 16.

47 Pickering, 2001.

48 Mulvaney, 1991, 17.

49 Pickering, 2001.

50 Pickering, 2001.

51 Colley, 2002, 164–165. For examples, see Anon., May 1999a and b; Bowler, 2002 (although note Bowler's awareness of the discomfort that the images might cause Indigenous people, and the warning at the start of the CD); Grose, 2000, 3; Thorne *et al.*, 1999, 592.

52 Pickering, 2001.

53 Bowler, 2002; Mulvaney, 1991, 17.

54 Bowler, 2002.

55 Victorian Quaternary Group Papers, Circular 30, 8 October 1981 (private collection of E.B. Joyce, School of Earth Sciences, University of Melbourne). Bowler worked closely with Aboriginal people while doing geomorphological work near Noonkanbah, in Western Australia's Kimberley region, in the late 1970s.

56 Quoted in Jopson, 1999, 5. See also Colley, 2002, 165.

57 J.M. Bowler, pers. comm., 22 July 1999, Melbourne.

'The desired future'

1 Bowler and Magee, [1996?], 1.

2 Manidis Roberts Consultants, 1996, 9.

3 Bowler, 1976, 279–310; Bowler and Jones, 1979.

Afterword

1 O'Riordan, Shadrake and Wood, 1989, 184.

2 Seddon, 1996, 495.

3 Ride and Davis, 1997, 197–221. See also Rudwick, 1976.

4 Rudwick, 1997, 18–24; Lyell, 1830–1833; Agassiz, 1967.

5 Bowler, 2002.

6 Levell and Wee, 2009, 46; Anon., 24 May 1893, 4–5.

7 Seddon, 1996, 487, 493; Bowler 1995.

8 Laseron 1953; Laseron, 1954; Schama, 1995, 15.

9 Hans Mincham, pers. comm., 2 February 2000, Adelaide.

10 Bruce, 1922; Parker, 1953; Ramsay Smith, 1930; Reed, 1965.

11 Hutton and Connors, 1999, 109.

12 Bonyhady, 1996, 147; Davison, 2000, 113; Tom Uren quoted in Hughes, 1996.

13 Hall, 1992, 167.

References

Unpublished material

Allen, H.R., 1972. Where the crow flies backward: Man and land in the Darling basin. Unpublished PhD thesis. Australian National University, Canberra.

Anon., no date. Record of Mines Summary Card, No. 156, Hallett's Mine (Hallett Cove Papers).

Anon., 1980. Worthing Mine or Hallett Mine, Field River, Hallett Cove: Items No. 6628–10545. South Australian Register of State Heritage Items, Base Data Sheet, 24 July 1980.

Australian Government, 1980. Nomination of the Willandra Lakes Region for inclusion in the World Heritage List. Australian Government, Canberra.

Bonython, C.W., G.W. Toogood, E.M. McBriar and I.M. Thomas, 1975. Submission to committee appointed by the Australian government to enquire into the preservation of Hallett Cove (Rudman Report). Unpublished report, National Trust of South Australia (Hallett Cove Papers).

[Booth, E.S.], no date. unpublished TS, Stirling Papers, AA 309, South Australian Museum, Adelaide.

Bowler, J.M. and J.W. Magee, [1996?]. Geomorphology of the Willandra Lakes Region World Heritage Area. Unpublished report to the Willandra Lakes World Heritage Plan of Management Committee, Australian Government, Canberra.

Bowler, J.M. and M.L. Smith, 1996. Willandra Lakes Region World Heritage Area excursion guide: Introductory notes, Museum of Victoria teachers' tour, 5–7 July 1996. Unpublished TS.

Callabonna Papers, correspondence, notes, minutes, cuttings, sketches and miscellaneous MSS and TSS relating to expeditions to the Lake Callabonna fossil reserve in the Department of Vertebrate Palaeontology, South Australian Museum (hereafter Callabonna Papers), Adelaide.

Clarke, J., 1975. Hallett Cove glacial pavements: notes on preservation negotiations from 5th April 1957 to 6th December 1965. Unpublished report for the National Trust of South Australia, 5 August 1975 (private collection of C.W. Bonython).

Clarke, William Branwhite, Papers, State Library of New South Wales, Sydney.

Connell, J.E., 1982. Worthing Mine, Hallett Cove. Unpublished report, Department of Environment and Planning and Department of Mines and Energy, SA, 11 August 1982 (Hallett Cove Papers).

[Cooper, B.J.], 1977. Nomination of Hallett Cove to the National Estate on behalf of the Geological Society of South Australia. Unpublished MS, 21 April 1977 (private collection of B.J. Cooper).

Cooper, B.J., 1992. Pioneering geologists and Hallett Cove. Unpublished TS of paper delivered to the Friends of Hallett Cove, 14 February 1992 (private collection of B.J. Cooper).

Darwin Papers (DAR), Cambridge University Library.

David, T.W.E., 1928. Correspondence, T.W.E. David (Sydney) to Professor J.W. Gregory, President, Geological Society (Glasgow, Scotland), 20 July 1928, Edgeworth David Papers, University of Sydney Archives, Series 16.

Department of Environment and Conservation, NSW (DEC), 2006. Mungo National Park plan of management. National Parks and Wildlife Service, NSW Government, Sydney. Online <www.environment.nsw.gov.au/parkmanagement/ParkManagementPlans.htm>.

Department of the Environment, Heritage and Aboriginal Affairs, South Australia (DEHAA), 1998. Annual report 1997–1998. South Australian Government, Adelaide.

DEHAA, 1999. Annual report 1998–1999. South Australian Government, Adelaide.

Department of the Environment and Natural Resources, South Australia (DENR), 1994. Review into the management of the National Parks and Wildlife Act: Final Report. South Australian Government, Adelaide.

DENR, 1997. Annual report 1996–1997. South Australian Government, Adelaide.

Douglas, Kirsty, 1996. Land systems and stratigraphy of Lake Mulurulu: Examination of Quaternary palaeoenvironments. Unpublished Honours report, University of Melbourne.

Douglas, Kirsty, 1998a. The roving geographer: Reading space, place and inversion in colonial Australian landscapes. Unpublished MPhil thesis, University of Cambridge.

Douglas, Kirsty, 2004. Pictures of time beneath: Science, landscape, heritage and the uses of the deep past in Australia, 1830–2003. Unpublished PhD thesis. Australian National University, Canberra.

Evans, E., T. Smith, P. Bell, R. Hickman, M. Schulze, W. Coomans, N. Hiern, J. Fenton, J. Treloar, B. Kavanagh, C. Barnes and P. Whiffen, 1986. Worthing Mine and environs report. Report prepared by the Worthing Mine and Environs Working Group for the Minister for Environment and Planning, SA (Hallett Cove Papers).

Ellis, R., and G. Warnes, 1977. Hallett Cove Conservation Park, an environmental resource. Unpublished student report, Sturt CAE, Environmental Studies (Hallett Cove Papers).

Fletcher, Harold O., 1996. Delving into the past: Incidents in the career of a palaeontologist and a brief introduction to interesting fossils of Australia. Unpublished TS, Australian Museum Archive ref. 96/14P, Sydney.

Gill, Edmund D., Papers, PA884, Notebooks 1–58. La Trobe Collection, State Library of Victoria, Melbourne.

Hallett Cove Papers. Correspondence, photographs and miscellaneous MSS and TSS relating to Hallett Cove and geological heritage in South Australia, 1960–1998, private collection of E.M. McBriar, Adelaide.

Hamilton, C., 1972. Subdividing Halletts Cove. Unpublished Environmental Law seminar paper, University of Adelaide Law School (Hallett Cove Papers).

Hobson, Edmund Charles, Papers, MS8457. La Trobe Collection, State Library of Victoria, Melbourne.

Hobson, E.C., 1839. 'Diary of a journey with Lady Franklin's party overland from Melbourne to the Hume River', Box 25/1. Hobson Papers, MS383/09. La Trobe Collection, State Library of Victoria, Melbourne.

Howchin, W., [1911?]. Mr H. Basedow and the glacial beds of Cambrian age in South Australia. Unpublished MS, MSS 9, series 1.1, W. Howchin, The Basedow–Howchin controversy 1909–1911. In Sir D. Mawson, On South Australian geology, Barr Smith Library Special Collections, University of Adelaide.

Hurst, Henry, 1893. Preliminary report from H. Hurst to the Museum Committee, 29 July 1893 (Callabonna Papers).

Joyce E.B. and R.L. King, 1980. Geological features of the National Estate in Victoria: An inventory compiled for the Australian Heritage Commission (Geological Society of Australia, Victorian Division: Melbourne).

Joyce, E.B., 1995. Assessing the significance of geological heritage: A methodology study for the Australian Heritage Commission. Report prepared for the Australian Heritage Commission by the Standing Committee for Geological Heritage, GSA (Geological Society of Australia: Melbourne).

Krefft, Gerard, 1861–1878. Correspondence, Albert Gunther papers. Unpublished TS, Australian Museum Archive ref. AMS48, Sydney.

McBriar, E.M., 1968. Notes on Hallett Cove for Mr Hiles, secretary of the National Trust. Unpublished TS, 8 November 1968 (Hallett Cove Papers).

McBriar, E.M., 1971. A brief note on the conservation of Hallett Cove, South Australia. Unpublished TS, 16 June 1971 (Hallett Cove Papers).

McBriar, E.M., 1975. Submission to the Committee of Enquiry into the conservation of Hallett Cove (Rudman Report), May 1975. Unpublished report, South Australian Division of the Geological Society of Australia, 1 May 1975 (Hallett Cove Papers).

McDonald, I.N., *et al.*, 1965. The conservation of Hallett Cove. Unpublished student report, Louis Laybourne Smith School of Architecture and Building, Department of Planning and Landscape, South Australian Institute of Technology (Hallett Cove Papers).

Macquarie, Lachlan, 1816–1818. Diary of Governor Lachlan Macquarie, Macquarie Papers, Mitchell Library ML CYA 773, Sydney.

Martin, T.S. and Associates, 1971. Hallett Cove development application. Unpublished report prepared for Kadima Pty Ltd and Silesia Pty Ltd, Perth, August 1971 (Hallett Cove Papers).

Mitchell, T.L., *c.* 1801–1849. Papers of Sir T.L. Mitchell, vol. 8, C9A 295, State Library of New South Wales, Sydney.

Mount, T.J., 1975. Map of Hallett Cove Conservation Park: Recommended developments for site management. Annotated map accompanying unpublished report commissioned by the South Australian Coast Protection Board, November 1975 (Hallett Cove Papers).

Murchison, R., Correspondence of Sir Roderick Murchison, LDGSL 838. Geological Society Library, London (Murchison Papers).

Nancarrow, H.F., 1971. Conservation of geological features: Hallett Cove project. Unpublished report, Hallett Cove Development Company Pty. Ltd., Adelaide, 6 April 1971 (Hallett Cove Papers).

Owen, Richard, Letters, Richard Owen Papers, Natural History Museum, London (Owen Papers).

Pickering, Michael, 2001. Repatriation, rhetoric and reality: The repatriation of Australian Indigenous human remains and sacred objects. Paper presented at the Australian Registrars Committee Conference, 9 October 2001.

Player, Ann, 1990. Julian Tenison Woods 1832–1889: The interaction of science and religion. Unpublished MA thesis, Australian National University, Canberra.

Ragless, F.B., no date. Seventy years ago: The journal of Frederick Brandis Ragless. Unpublished MS (private collection of Margaret E. Ragless).

Rudman, P., P.W. Cullen, K. Enright, C.D. Ollier and R.W. Robertson, 1975. Hallett Cove study report: An investigation of the scientific, environmental and social issues relating to urban development in the Hallett Cove area. Report commissioned under the National Estate Program, for the Minister for Urban and Regional Development, Mr Tom Uren. Australian Government, Canberra (Hallett Cove Papers). (Rudman Report).

Sibly, J., R. Bauer, C. Burgess, G. Croser, J. Genimahaliotis, V. Peganite and K. Trimper, 1975. Planning for Hallett Cove: A report. Unpublished report by the Adelaide College of Adult Education commissioned by the Corporation of the City of Marion (Hallett Cove Papers).

South Australian Science Teachers Association (SASTA), 1971. The need for the Government to acquire and preserve an extensive area of land at Hallett Cove as Major District Open Space. Unpublished report (private collection of C.W. Bonython).

Stirling, E.C., Stirling Papers, Anthropology Archive (AA) 309, South Australian Museum, Adelaide (henceforth AA 309, SAM, Adelaide).

Stirling, Edward Charles, 1848–1919. PRG 388 Stirling Family Papers (compiled E.S. Booth), Mortlock Collection, State Library of South Australia (henceforth Booth Papers, SLSA).

Stirton, R.A., 1953. Field journal, February 12, 1953 to August 23, 1953, pp. 1–263, unpublished TS (UCMP Collection).

Stirton, R.A., 1954. Notebook 1, Australia, unpublished TS (UCMP Collection).

Stirton, R.A., 1954–55. Letterbook copies, Correspondence Box 1 (correspondence, UCMP Collection).

Stirton, R.A., 1958. Field journal 1958, South Australia, unpublished TS (UCMP Collection).

Swart, Rosemary, 1992. Environmental protection of Geological Monuments in South Australia. Unpublished MSc thesis, University of Adelaide.

Szent-Ivany, J.J.H., 1975. Importance of the preservation of Hallett Cove as a national reserve from the entomologist's and insect conservationist's point of view. Unpublished report for the Nature Preservation Committee of the National Trust of South Australia (Hallett Cove Papers).

Tate, R.L., [1894?–1901]. Elementary geology lectures: notes and sketches. Unpublished notebooks, Ralph Tate – Notebooks, 2 vol., MSS 0016, Barr Smith Library Special Collections, University of Adelaide.

Taylor, Norman, 1854?–1894. Personal file of newspaper cuttings and notes, MSB515. La Trobe Collection, State Library of Victoria, Melbourne.

Taylor, N., 1873–1875. Field book of Norman Taylor, Geologist and field surveyor, Geological Survey of Victoria [interleaved with A.R.C. Selwyn and G.H.F. Ulrich, 1866, *Notes on the Physical Geography,*

Geology and Mineralogy of Victoria], MS6216, La Trobe Collection, State Library of Victoria, Melbourne, Box 239/3.
Tedford, Richard H., 1953. Field journal, Australia expedition, 1953, part 1, 15 March – 29 June. Unpublished MS, private collection of Richard H. Tedford.
Tedford, Richard H., 1953. Field journal, Australia expedition, 1953, part 3, 23 July – 29 October. Unpublished MS (UCMP Collection).
UCMP Collection. Correspondence, field journals, sketches and miscellaneous MSS and TSS, University of California Museum of Paleontology, University of California, Berkeley.
Victorian Quaternary Group Papers. Private collection of E.B. Joyce, School of Earth Sciences, University of Melbourne.
[de Vis, C.W.], 1889. Annual Report of the Curator. In Annual Report of the Trustees of the Queensland Museum, CA 17 – 1889 (Callabonna Papers).
Walkom, A.B., 1945. Memo to the Secretary, Liquid Fuel Control Board, Sydney, 30 May 1945. Australian Museum Correspondence Item 35/43, Sydney.
Wallace, A.R., Alfred Russel Wallace Papers, 1064B, Add. MS 46435, British Library Manuscripts, British Library, London.
Whitten, G.F., 1967. The preservation of Hallett Cove as a national geological monument. South Australian Department of Mines Report Book No. 64/50, Geological Survey, Exploration Services Division, 23 March 1967.

Published material

Agassiz, E.C., ed., 1886. *Louis Agassiz, his Life and Correspondence,* 2 vols (Houghton, Mifflin and Company: Boston).
Agassiz, Jean Louis Rodolphe, 1967. *Studies on Glaciers, Preceded by the Discourse of Neuchâtel,* ed. and tr. A.V. Carozzi (Hafner Publishing: New York).
Agassiz, L., 1840. *Etudes sur les Glaciers,* 8 vols (Neuchâtel, Switzerland).
Agassiz, Louis, 1866. *Geological Sketches* (Ticknor and Fields: Boston).
Aitken, M.J., 1998. *An Introduction to Optical Dating: The Dating of Quaternary Sediments by the Use of Photon-stimulated Luminescence* (Oxford University Press: Oxford).
Alderman, A.R., 1967. The development of geology in South Australia: A personal view, *Records of the Australian Academy of Science* 1(2): 29–51.
Allen, H.R., 1974. The Bagunji of the Darling Basin, *World Archaeology* 5: 309–322.
Allen, Harry, 1998. Reinterpreting the 1969–1972 Willandra Lakes archaeological surveys, *Archaeology in Oceania* 33(3), Special Issue: 207–220.
Allen, Harry, Simon Holdaway, Patricia Fanning and Judith Littleton, 2008. Footprints in the sand: Appraising the archaeology of the Willandra Lakes, western New South Wales, Australia, *Antiquity* 82(315): 11–25.
Allen, Harry and Simon Holdaway, 2009. The archaeology of Mungo and the Willandra Lakes: looking back, looking forward, in *25 Years of World Heritage Status for Willandra Lakes: Papers from the Legacy of an Ice Age Commemorative Conference, 2006*, eds Arthur Durband, Dora Constantinidis and Daniel Rayner, *Archaeology in Oceania* 44: 74–118, pp. 96–106.
Andrews, Alan E.J., ed., 1986. *Stapylton: With Major Mitchell's Australia Felix Expedition, 1836, Largely from the Journal of Granville William Chetwynd Stapylton* (Blubber Head Press: Hobart).
Anon., 1884. Third excursion, New Year's Day, *Transactions of the Royal Society of South Australia* 7: 116–118.
Anon., 1886. Thirteenth excursion, Tuesday, September 1, 1885: Hallett's Cove, *Transactions of the Royal Society of South Australia* 9: 192–193.
Anon., 1893. The discovery of fossil remains at Lake Mulligan, *South Australian Register,* 23 May.
Anon., 1893. Extinct animals of Australia, *South Australian Register,* 24 May.
Anon., 1893. The discovery of fossil remains at Lake Mulligan, *Adelaide Observer,* 27 May.
Anon. [H.Y.L. Brown], 1893. Complaint about state of the fossils, *South Australian Register,* 20 June: 4.

Anon., 1893. Arrival of mammoth fossils from Mulligan Lake, *South Australian Register,* 31 July.

Anon., 1894. The fossils from Lake Mulligan, *South Australian Register,* 12 September.

Anon., 1895. Glacial features of Hallett's Cove, report of a paper read by W. Howchin to the Royal Society of South Australia, 3 April 1895, *South Australian Register,* 6 April.

Anon., 1919. Obituary notice for E.C. Stirling, *Transactions of the Royal Society of South Australia* 43: 1–2.

Anon., 1966. Proposed geological reservation at Halletts Cove, *Fossicker* June: 7–8.

Anon., 1971. Developer alters Hallett Cove plan, *Advertiser* (Adelaide), 15 September.

Anon., 1971. 'Relax' is a foreign word, *News* (Adelaide), 20 September.

Anon., 1999a. Press release, The dating of an ancient skeleton raises questions about human evolution in Australia, embargoed to 20 May 1999, Australian National University, Canberra.

Anon., 1999b. New dates for Mungo III announced, *ANU Archaeology Online Newsletter,* May.

Anon., 2003. Making tracks: Aboriginal tours at Lake Mungo, *Age* travel supplement, 16 August: 6.

Aplin, Graeme, 2002. *Heritage: Identification, Conservation, and Management* (Oxford University Press: Melbourne).

Appleton, Josie, 2002. Battle of the bones, *Spiked Culture* 12 December. Online <www.spiked-online.com/Articles>.

Aurousseau, M., ed. and tr., 1968. *The Letters of Ludwig Leichhardt,* 3 vols (Cambridge University Press, Hakluyt Society: Cambridge, UK).

Austin, J.B., 1863. *The Mines of South Australia* (Facsimile reprint: State Library of South Australia: Adelaide 1968).

Austin, P., 1990. The last words of Pirlatapa, *Pacific Linguistics C* (116), *Landscape and History: Essays in honour of Luise A. Hercus,* 29–48.

Babbage, B.H., P.E. Warburton and A.P. Burtt, 1858. Northern exploration. Reports from Messrs Babbage and Warburton and police trooper Burtt on exploration into the north and north-western interior of South Australia, *Parliamentary Papers of South Australia* 151 (Government Printer: Adelaide), 1–19.

Banks, Joseph, 1962. *The* Endeavour *Journal of Joseph Banks, 1768–1771,* ed. J.C. Beaglehole, 2 vols (Public Library of New South Wales with Angus & Robertson: Sydney).

Barnes, Jill, 2001. Hoges works wonders for the Land-down-under: A nostalgic chapter in Australian tourism marketing, *Melbourne Historical Journal: Proceedings of the Mass Historia Postgraduate Conference, University of Melbourne* 29: 29–36.

Barthes, Roland, 1993 (1957). *Mythologies,* tr. Annette Lavers (Vintage: London).

Basalla, George, 1967. The spread of western science, *Science* 156(3775) May: 611–622.

Batty, Philip, 1998. Saluting the dot-spangled banner: Aboriginal culture, national identity and the Australian republic, *Australian Humanities Review* September. Online <www.lib.latrobe.edu.au/AHR/archae/Issue-September-1998/batty.html>

de Beer, G., 1960. Darwin's notebooks on transmutation of species, Part II, second notebook (July 1837 to February 1838), *British Museum (Natural History) Historical Series* 2(3).

Bednall, W.T., 1878. Australian Trigonias and their distribution, *Transactions and Proceedings and Report of the Philosophical Society of Adelaide, South Australia* 1.

Benedict, Jeff, 2003. *No Bone Unturned: The Adventures of a Top Smithsonian Forensic Scientist and the Legal Battle for America's Oldest Skeleton* (HarperCollins: New York).

Benjamin, Walter, 1969. *Illuminations,* ed. Hannah Arendt, tr. Harry Zohn (Schocken: New York).

Black, E.C., 1962. The Lake Torrens hoodoo, *Proceedings of the Royal Geographical Society of Australasia* (South Australian Branch) 64: 43–51.

Bolles, E.B., 1999. *The Ice Finders: How a Poet, a Professor, and a Politician Discovered the Ice Age* (Counterpoint: Washington DC).

Bonyhady, Tim, 1996. The stuff of heritage, in *Prehistory to Politics: John Mulvaney, the Humanities and the Public Intellectual,* eds T. Bonyhady and T. Griffiths (Melbourne University Press: Melbourne), 144–162.

Bonyhady, Tim, 1997. Disturbing the dead, *Art Monthly Australia,* November: 9–12.

Bonython, C.W., 1975. Conservation of the coast, *The National Trust of South Australia Newsletter* 72: 3.

Bonython, C.W., 1975. Our role in conserving Hallett Cove, *The National Trust of South Australia Newsletter* 74: 3.

Bourman, R.P. and N.F. Alley, 1990. Stratigraphy and environments of deposition at Hallett Cove during the Late Palaeozoic, *Mines and Energy Review, South Australia* 157: 68–83.

Bourman, R.P. and N.F. Alley, 1995. Troubridge Basin, in *The Geology of South Australia, Vol. 2: The Phanerozoic*, eds J.F. Drexel and W.V. Preiss (Geological Survey of South Australia: Adelaide), 65–70.

Bowler, J.M., 1971. Pleistocene salinities and climate change: Evidence from lakes and lunettes in south-eastern Australia, in *Aboriginal Man and Environment in Australia*, eds D.J. Mulvaney and J. Golson (ANU Press: Canberra), 47–65.

Bowler, J.M., 1976. Aridity in Australia: Age, origins and expression in aeolian landforms and sediments, *Earth Science Reviews* 12: 279–310.

Bowler, J.M., 1981. Australian salt lakes: A palaeohydrologic approach, *Hydrobiologia* 82: 431–444.

Bowler, J.M., 1982. Aridity in the late Tertiary and Quaternary of Australia, in *Evolution of the Flora and Fauna in Arid Australia*, eds W.R. Barker and P.J.M. Greenslade (Peacock Publications: Frewville), 35–44.

Bowler, J.M., 1983. Lunettes as indices of hydrologic change: A review of Australian evidence, *Proceedings of the Royal Society of Victoria* 95: 147–168.

Bowler, J.M., 1986. Quaternary landform evolution, in *The Natural Environment*, ed. D.N. Jeans, *Australia, a Geography*, vol. 1 (Sydney University Press: Sydney), 117–147.

Bowler, J.M., 1990. The last 500,000 years, in *The Murray*, eds N. Mackay and D. Eastburn (Canberra: Murray–Darling Basin Commission), 95–109.

Bowler, Jim, 1995. Reading the Australian landscape: European and Aboriginal perspectives, *Cappuccino Papers* 1, Imagine the future Inc, Melbourne. Online <www.ecoversity.org.au/publications/bowler1995.htm>.

Bowler, J.M., 1998. Willandra Lakes revisited: Environmental framework for human occupation, *Archaeology in Oceania* 33(3), Special issue: 120–155.

Bowler, J.M., 2000. Mungo dates out of kilter, *Australian,* opinion, 10 January: 11.

Bowler, J.M., 2002. *Lake Mungo: Window to Australia's past,* CD-ROM (University of Melbourne and Murray–Darling Basin Commission: Melbourne).

Bowler, Jim, 2005. John Mulvaney: An amazing journey. *UniNews* University of Melbourne 14(7), 2–16 May <uninews.unimelb.edu.au/news/2304/uninewsarchive.php?volume=14&number=7&publication=un>.

Bowler, J.M., R. Jones, H. Allen and A.G. Thorne, 1970. Pleistocene human remains from Australia: A living site and human cremation from Lake Mungo, western New South Wales, *World Archaeology* 2: 39–60.

Bowler, J.M., A.G. Thorne and H.A. Polach, 1972. Pleistocene man in Australia: Age and significance of the Mungo skeleton, *Nature* 240: 48–50.

Bowler, J.M., G.S. Hope, J.N. Jennings, G. Singh and D. Walker, 1976. Late Quaternary climates of Australia and New Guinea. *Quaternary Research* 6: 359–394.

Bowler, J.M. and J.W. Magee, 1978. Geomorphology of the Mallee region in semi-arid northern Victoria and western New South Wales, *Proceedings of the Royal Society of Victoria* 90: 5–25.

Bowler, J.M. and Rhys Jones, 1979. Australia was a land of lakes, *The Geographical Magazine* July: 679–685.

Bowler, J.M. and D.M. Price, 1998. Luminescence dates and stratigraphic analyses at Lake Mungo: Review and new perspectives, *Archaeology in Oceania* 33(3), Special issue: 156–168.

Bowler, J.M. and J.W. Magee, 2000. Redating Australia's oldest human remains: A sceptic's view, *Journal of Human Evolution* 38: 719–726.

Bowler, Peter J, 1976. *Fossils and Progress: Paleontology and the Idea of Progressive Evolution in the Nineteenth Century* (Science History Publications: New York).

Branagan, David, 1992. Richard Owen in the Antipodean context: a review, in Richard Owen, Thomas Mitchell and Australian science: A commemorative symposium, ed. David Branagan, *Journal of the Proceedings of the Royal Society of New South Wales* 125(3–4): 95–102.

Brown, C.M., 1989. Structural and stratigraphic framework of groundwater occurrence and surface discharge in the Murray Basin, southeastern Australia, *BMR Journal of Australian Geology and Geophysics* 11: 127–146.

Brown, C.M. and A.E. Stephenson, 1991. Geology of the Murray Basin, southeastern Australia, *BMR Geology and Geophysics Bulletin* 235.

Brown, H.Y.L., 1892. Government geologist's report on country in neighbourhood of Lake Eyre, *Parliamentary Papers of South Australia,* 141 (Government Printer: Adelaide).

Brown, H.Y.L., 1893. Fossil bones near Callabonna Station, *South Australian Register,* 1 July.

Brown, H.Y.L., 1894. Annual report of the government geologist for the year ended 30 June 1894, *Parliamentary Papers of South Australia,* 25 (Government Printer: Adelaide).

Brown, Peter, 1981. Artificial cranial deformation: A component in the variation in Pleistocene Australian crania, *Archaeology in Oceania* 16: 156–167.

Brown, Peter, 2000. Australian Pleistocene variation and the sex of Mungo 3, *Journal of Human Evolution* 38: 743–749.

Browne, Janet, 1983. *The Secular Ark: Studies in the History of Biogeography* (Yale University Press: New Haven).

Bruce, Mary Grant, 1922. *The Stone Axe of Burkamukk* (Ward, Lock & Co. Ltd: London and Melbourne).

Buckland, William, 1823. Reliquiae diluvianae, *or Observations on the Organic Remains Contained in Caves, Fissures, and Diluvial Gravel and on other Geological Phenomena Attesting the Action of a Universal Deluge* (John Murray: London).

Buckland, William, 1827. Observations on the bones of hyenas and other animals in the Cavern of Lunel, near Montpellier and in the adjacent strata of marine formation, *Edinburgh Journal of Science* 6: 242–246.

Buckland, William, 1831. Ossemons decouvertes a la Nouvelle Hollande, *Bulletin de la Société géologique de France* 1: 227.

Buckland, William, 1842. On the former existence of glaciers in Scotland and in the north of England, *Proceedings of the Geological Society, London* 3: 332–348.

Buffetaut, Eric, 1987. *A Short History of Vertebrate Palaeontology* (Croom Helm: London).

Butcher, Barry W., 1989. 'Adding stones to the great pile'? Charles Darwin's use of Australian resources, 1837–1882, *Historical Records of Australian Science* 8(1) December: 1–15.

Bynum, W.F., E.J. Browne and R. Porter, eds, 1983. *Dictionary of the History of Science* (Macmillan: London).

Byrne, Denis, 1996. Deep nation: Australia's acquisition of an indigenous past, *Aboriginal History* 20: 82–107.

Cadbury, Deborah, 2000. *The Dinosaur Hunters: A Story of Scientific Rivalry and the Discovery of the Prehistoric World* (Fourth Estate: London).

Caldicott, R., ed., 1971. Hallett Cove carve-up, *Planning SA, incorporating the South Australian Conservation Bulletin* 19, January/February: 1.

Caldicott, R. and A. Geering, 1974. Development and disaster: Hallett Cove, *Habitat,* June: 20–22.

Caldicott, R. and A. Geering, 1975. 'Operation Hallett', *Sunday Mail* (Adelaide), 4 May.

Cameron, A.L.P., 1885. Notes on some tribes of New South Wales, *Journal of the Royal Anthropological Institute of Great Britain and Ireland* 14(4): 344–370.

Carozzi, A.V., 1966. Agassiz's amazing geological speculation: The Ice Age, *Studies in Romanticism* 5(2): 57–83.

Carter, David, 1995. Future pasts, in *The Abundant Culture: Meaning and Significance in Everyday Australia,* eds David Headon, Joy Hooton and Donald Horne (Allen & Unwin: St Leonards), 3–15.

Carter, Paul, 1988. *The Road to Botany Bay: An Exploration of Landscape and History* (Knopf: New York).
Carter, Paul, 1992. *Living in a New Country: History, Travelling and Language* (Faber: London).
Cathcart, Michael, 2002. Uluru, in *Words for Country: Landscape and Language in Australia*, eds Tim Bonyhady and Tom Griffiths (University of New South Wales Press: Sydney), 206–221.
Chippindale, Christopher, 1985. Skeletons rattle down under, *New Scientist*, 14 March.
Clarke, W.B., 1847. Fossil bones, *Sydney Morning Herald*, 30 November.
Clarke, W.B., 1853. Report to the Government of NSW, *Parliamentary Papers of New South Wales.*
Clarke, W.B., 1860. *Researches in the Southern Goldfields of New South Wales* (Reading and Wellbank: Sydney).
Cockburn, R., 1984 (1908). *What's in a Name? Nomenclature of South Australia: Authoritative Derivations of some 4000 Historically Significant Place Names*, edited, revised and enlarged by Stuart Cockburn (Ferguson Publications: Adelaide).
Colhoun, E.A. and J.A. Peterson, 1986. Quaternary landform evolution and cryosphere: Research progress from Sahul to Australian Antarctica, *Australian Geographical Studies* 24: 145–167.
Colley, Sarah, 2002. *Uncovering Australia: Archaeology, Indigenous People and the Public* (Allen & Unwin: Sydney).
Colwell, M., 1972. *The History of the Noarlunga District* (Scrymgour and Sons: Adelaide).
Cooper, H.M., 1959. Large archaeological stone implements from Hallett Cove, South Australia, *Transactions of the Royal Society of South Australia* 82: 55–60.
Cooper, H.M., 1968a. A further description of large stone implements from South Australia, *Records of the South Australian Museum* 15: 581–603.
Cooper, H.M., 1968b. An archaeological camp-site in the vicinity of Hallett Cove, South Australia, *Records of the South Australian Museum* 15: 625–714.
Cooper, H.M., M. Kenny and J.M. Scrymgour, 1972 (1970) *Hallett Cove: A Field Guide,* 2nd edition (South Australian Museum: Adelaide).
Corbett, D.W.P., 1977. Profiles of famous Australian geologists: 1. Ralph Tate 1840–1901, *Bulletin of the Field Geology Club of SA* 6(8): 14–18.
Corbett, D.W.P., 1978. Profiles of famous Australian geologists: 3. Walter Howchin 1845–1937, *Bulletin of the Field Geology Club of SA* 7(3): 16–19.
Croll, J.W., 1864. On the physical cause of the change of climate during geological epochs, *Philosophical Magazine* 28: 121–137.
Croll, J.W., 1866. Physical cause of the submergence of the land during the glacial epoch, *Philosophical Magazine* 31: 301–305.
Croll, J.W., 1867. The change in the obliquity of the ecliptic, its influence on the climate of the polar regions and on the level of the sea, *Philosophical Magazine* 33: 426–445.
Croll, J.W., 1869. On the physical cause of the motion of glaciers, *Philosophical Magazine* 37: 201–208.
Croll, J.W., 1875. *Climate and Time in their Geological Relations* (Daldy, Isbister and Co: London).
Cumpston, J.H.L., 1972. *Augustus Gregory and the Inland Sea* (Roebuck Society Publications: Canberra).
Dampier, William, 1906a (1697, 6th ed., 1717). *A New Voyage Around the World,* vol. 1, in *Dampier's Voyages*, ed. John Masefield, vol. 1, 15–529 (E. Grant Richards: London).
Dampier, William, 1906b (1703). *A Voyage to New Holland &c. in the year 1699,* in *Dampier's Voyages,* ed. John Masefield, vol. 2, 451–573 (E. Grant Richards: London).
Daniel, G., ed., 1981. *Towards a History of Archaeology* (Thames and Hudson: London).
Darwin, Charles, 1842. *On the Structure and Distribution of Coral Reefs. Being the First Part of the Geology of the Voyage of the Beagle under the Command of Capt FitzRoy, RN, During the Years 1832 to 1836* (Smith, Elder and Co: London).
Darwin, Charles, 1844, *Geological Observations on the Volcanic Islands and Parts of South America Visited during the Voyage of HMS Beagle, Together with some Brief Notices of the Geology of Australia and the Cape of Good Hope. Being the Second Part of the Geology of the Voyage of the Beagle under the Command of Capt FitzRoy, RN, During the Years 1832 to 1836* (Smith, Elder and Co: London).

Darwin, Charles, 1845 (1839). *Journal of Researches into the Natural History and Geology of the Countries Visited by HMS Beagle under the Command of Capt FitzRoy, RN*, 2nd edition, corrected, with additions (John Murray: London).

Darwin, Charles, 1861 (1859). *On the Origin of Species by Means of Natural Selection, or the Preservation of Favoured Races in the Struggle for Life*, 3rd edition (John Murray: London).

David, T.W.E., 1893. 'Report of the research committee appointed to collect evidence as to glacial action in Australasia: Australia and Tasmania', *Report of the Fifth Meeting of the Australasian Association for the Advancement of Science*, Adelaide, 229–232.

David, T.W.E., 1895. 'Evidences of glacial action in Australia and Tasmania', president's address, Section C (Geology and Mineralogy), *Report of the Sixth Meeting of the Australasian Association for the Advancement of Science*, Brisbane, 59–98.

David, T.W.E., 1925. Response to D. Campbell (26 February 1925), *Sydney Morning Herald*, 4 March.

Davis, Richard C., ed., 2002. *The Central Australian Expedition, 1844–1846: The Journals of Charles Sturt*, Third Series, No. 10 (Hakluyt Society: London).

Davison, Graeme, 2000. *The Use and Abuse of Australian History* (Allen & Unwin: Sydney).

Dawson, A.G., 1992. *Ice Age Earth: Late Quaternary Geology and Climate* (Routledge: London).

Dillon, Rodney, 2004. Show ancient bones proper respect, *New Scientist*, 14 February.

Dolling, Alison, 1981. *The History of Marion on the Sturt: The Story of a Changing Landscape and its People* (Peacock Publications: Frewville).

Douglas, Bronwen, 1998b. *Across the Great Divide: Journeys in History and Anthropology* (Harwood Academic Press: Amsterdam).

Douglas, Bronwen and Chris Ballard, ed., 2008. *Foreign Bodies: Oceania and the Science of Race 1750–1940* (ANU E Press: Canberra).

Douglas, J.G. and J.A. Ferguson, ed., 1988. *Geology of Victoria*, Geological Society of Australia special publication 5.

Douglas, Kirsty, 2001. The Diprotodon's toe: Serendipity, parochialism and the mobilisation of the deep past in colonial South Australia, *Melbourne Historical Journal: Proceedings of the Mass Historia Postgraduate Conference, University of Melbourne 2001*, 29, 57–63.

Douglas, Kirsty, 2002. Scarcely any water on its surface, in *Words for Country: Landscape and Language in Australia*, eds Tim Bonyhady and Tom Griffiths (University of New South Wales Press: Sydney), 68–83.

Douglas, Kirsty, 2006. 'Forsaken spot' to 'classic ground': Geological heritage and the recuperative power of the deep past, *Environment and History* 12(3): 269–296.

Domin, Eduard R., Hans Mincham, Robert Swinburne and Jean Cook, 1986. *The Flinders Ranges: A Portrait* (Little Hills Press: Adelaide).

Duffy, Patrick K. and Lois A. Lofgren, 1994. Jurassic farce: A critical analysis of the Government's seizure of 'Sue'™, a sixty-five-million-year-old Tyrannosaurus Rex fossil, *South Dakota Law Review* 39(3): 478–528.

Dugan, Kathleen G., (1980) 1979. Darwin and *Diprotodon*: the Wellington Caves fossils and the law of succession, *Proceedings of the Linnean Society of New South Wales* 104(4): 265–272.

Edwards, Robert, ed., 1975. *The Preservation of Australia's Aboriginal Heritage* (Australian Institute of Aboriginal Studies: Canberra).

Egan, Timothy, 2000. US takes tribes' side on bones, *New York Times*, 26 September.

Ellery, R.L.J., 1868. President's address, *Transactions and Proceedings of the Royal Society of Victoria* 9(1): 1–21.

Errey, K. and T. Flannery, 1978. The neglected megafaunal sites of the Colongulac region, western Victoria, *The Artefact* 3(2): 101–106.

Etheridge, R., Jr, 1896. Has man a geological history in Australia?, *Proceedings of the Linnean Society of New South Wales*, 2(5): 259–266.

Eyre, Edward John, 1845. *Journals of Expeditions of Discovery into Central Australia, and Overland from Adelaide to King George's Sound, in the years 1840–1; Sent by the Colonists of South Australia with the Sanction and Support of the Government: Including an Account of the Manners and Customs of the*

Aborigines and the State of their Relations with Europeans, 2 vols and maps (T. & W. Boone: London. Facsimile reprint, Libraries Board of South Australia: Adelaide, 1964).

Eyre, Edward John, 1984. *Autobiographical Narrative of Residence and Exploration in Australia 1832–1839*, ed. Jill Waterhouse (Caliban Books: London).

Fabian, Johannes, 1983. *Time and the Other: How Anthropology Makes its Object* (Columbia University Press: New York).

Falconer, Hugh, 1863. On the American fossil elephant of the regions bordering the Gulf of Mexico (*E. colombia*, Falc.); With general observations on the living and extinct species, *Natural History Review* 10: 43–114.

Fenner, C., 1937. Walter Howchin 1845–1937, obituary notice, *Transactions of the Royal Society of South Australia* 61: v–ix.

Fforde, Cressida, Jane Hubert and Paul Turnbull, 2002. *The Dead and their Possessions: Repatriation in Principle, Policy and Practice* (Routledge: London).

Field, Barron, ed., 1825. *Geographical Memoirs on New South Wales*, by various hands (John Murray: London).

Fiffer, Steve, 2000. *Tyrannosaurus Sue: The Extraordinary Saga of the Largest, Most Fought Over T. Rex Ever Found* (W.H. Freeman: New York).

Finlayson, H.H., 1935. *The Red Centre: Man and Beast in the Heart of Australia* (Angus & Robertson: Sydney).

Fischer, A.G., 1984. The two Phanerozoic supercycles, in *Catastrophes and Earth History*, eds W.A. Berggren and J.A. van Couvering (Princeton University Press: Princeton), 129–150.

Flannery, T.F., 1994. *The Future Eaters: An Ecological History of the Australasian Lands and People* (Reed New Holland Press: Sydney).

Flannery, Timothy Fridtjof, 2001. *The Eternal Frontier: An Ecological History of North America and its Peoples* (Text Publishing Company: Melbourne).

Flood, Josephine, 1983. *Archaeology of the Dreamtime: The Story of Prehistoric Australia and her People* (Collins: Sydney).

Flower, W.H., 1868. On the affinities and probable habitats of the extinct Australian marsupial *Thylacoleo carnifex* Owen, *Quarterly Journal of the Geological Society* 24: 307–319.

Forbes, B.J., 1994. South Australian Division, in *Rock Me Hard ... Rock Me Soft ... A History of the Geological Society of Australia Incorporated*, eds B.J. Cooper and D.F. Branagan (Sydney: Geological Society of Australia), 63–69.

Friends of Hallett Cove, 1999. Friends of Hallett Cove Conservation Park Membership Form and Sunday Programme.

Friends of the South Australian Museum (FOSAM), 1996. A visit to Hallett Cove in search of evidence of an ancient ice age, *Newsletter*, 19 October.

Frome, E.C., 1901. Extracts from the *Government Gazettes* of 1843 and 1844, cited in Appendix A, Letters from E.J. Eyre and enclosures thereto, *Proceedings of the Royal Geographical Society of Australasia (South Australian Branch)* 4, 1898–99 to 1900–01, 132–136.

Furphy, Sam, 2001. Aboriginal place-names and settler Australian identity, *Melbourne Historical Journal: Proceedings of the Mass Historia Postgraduate Conference*, University of Melbourne 2001, 29, 71–78.

Gee, Henry, 2000. *Deep time: Cladistics, the Revolution in Evolution* (HarperCollins: London).

Geikie, James, 1877. *The Great Ice Age and its Relation to the Antiquity of Man* (Edward Stanford: London).

Giesecke, R., ed., 1999. *A Field Guide to the Geology of Hallett Cove and Other Localities with Glacial Geology on the Fleurieu Peninsula* (Field Geology Club of South Australia: Adelaide).

Gill, E.D., 1953. Geological evidence in western Victoria relative to the antiquity of the Australian Aborigines, *Memoirs of the National Museum of Victoria* 18: 25–92.

Gillespie, Richard and Richard G. Roberts, 2000. On the reliability of age estimates for human remains at Lake Mungo, *Journal of Human Evolution* 38: 727–732.

Godthelp, H., 2001. Murgon site, *Riversleigh Notes: Australia's Lost Kingdoms* special edition 45: 5–7.

Gould, Stephen Jay, 1989. *Wonderful Life: The Burgess Shale and the Nature of History* (W.W. Norton: New York).
Gould, Stephen Jay, 1996. *Life's Grandeur: The Spread of Excellence from Plato to Darwin* (Random House: London).
Goyder, G.M., 1857. Northern exploration. Letter from the Hon. the Surveyor-General, enclosing report from Mr Goyder of his expedition into the northern districts of the colony, *Parliamentary Papers of South Australia* 72 (Government Printer: Adelaide), 1–4.
Grayson, Donald K., 1983. *The Establishment of Human Antiquity* (Academic Press: New York).
Grayson, Donald K., 1984. Nineteenth-century explanations of Pleistocene extinctions: A review and analysis, in *Quaternary Extinctions: A Prehistoric Revolution*, eds Paul S. Martin and Richard G. Klein (University of Arizona Press, Tucson) pp. 5–39.
Greenblatt, Stephen, 1988. *Marvellous Possessions: The Wonder of the New World* (Oxford University Press: Oxford).
Gregory, A.C., 1895. Presidential address: The geographical history of the Australian continent during its successive phases of geological development, *Report of the Sixth Meeting of the Australasian Association for the Advancement of Science*, Brisbane: 1–12.
Gregory, A.C. and F.T. Gregory, 1884. *Journals of Australian Explorations* (James C. Beal, Government Printer: Brisbane. Facsimile reprint, Libraries Board of South Australia: Adelaide, 1969).
Gregory, J.W., 1906. *The Dead Heart of Australia: A Journey around Lake Eyre in the Summer of 1901–1902, with Some Account of the Lake Eyre Basin and the Flowing Wells of Central Australia* (John Murray: London).
Gribbin, John and Mary Gribbin, 2001. *Ice Age* (London: Penguin Press).
Griffiths, G.S., 1884. On the evidence of the glacial epoch in Victoria during post-Miocene times, *Transactions of the Royal Society of Victoria* 21: 7–10.
Griffiths, Tom, 1996. *Hunters and Collectors: The Antiquarian Imagination in Australia* (Cambridge University Press: Melbourne).
Griffiths, Tom, 2001. Deep time and Australian history, *History Today* 51(11): 20–28.
Grose, Simon, 1999. Mungo 3 time frame dates to 68,000 years, *Canberra Times*, 21 May 1999.
Grose, Simon, 2000. Mungo Man critics say skeleton may belong to younger Mungo Girl, *Canberra Times*, 29 April 2000.
Grün, Rainer, Nigel Spooner, Alan Thorne, Graham Mortimer, John J. Simpson, Malcolm T. McCulloch, Lois Taylor and Darren Curnoe, 2000. Age of the Lake Mungo 3 skeleton, reply to Bowler & Magee and to Gillespie & Roberts, News and views, *Journal of Human Evolution* 38: 733–741.
Guy, Sandy, 2003a. Mungo jumbo, *Sunday Age* (Melbourne), 8 June 2003, future lifestyles supplement: 6.
Guy, Sandy, 2003b. Dry as a bone, *The Australian Way* Qantas in flight magazine, November 2003: 145–150.
Haast, J., 1867. Notes on Rev. J.T. Woods' paper 'On the glacial period in Australia', *Transactions and Proceedings of the Royal Society of Victoria* 8: 273–278.
Hall, C. Michael, 1992. *Wasteland to World Heritage: Preserving Australia's Wilderness* (Melbourne University Press: Melbourne).
Hallam, A., 1983. *Great Geological Controversies* (Oxford University Press: Oxford).
Hand, Susan and Michael Archer, eds, 1987. *The Antipodean Ark* (Angus & Robertson: Sydney).
Harley, M., 1994. The RIGS (Regionally Important Geological/geomorphological Sites) challenge: Involving local volunteers in conserving England's geological heritage, in *Geological and Landscape Conservation*, eds D. O'Halloran, C. Green, M. Harley, M. Stanley and J. Knill (London: Geological Society), 313–317.
Harris, Trudy, 2001. Mungo mad: search for the origin of man, Weekend focus, *Weekend Australian*, 13–14 January.
Hasenhohr, P., 1991/2. Pathways to the past: Fleurieu Peninsula: The geological heritage of South Australia, brochure (Geological Society of Australia Incorporated, South Australian Division: Adelaide).

Hill, Ernestine, 1981 (1951). *The Territory* (Angus & Robertson: Sydney).
Hills, E.S., 1940. The lunette: A new landform of aeolian origin, *Australian Geographer* 3(7): 15–21.
Hobson, E.C., 1845. Extract from a letter 'On some fossil bones discovered at Mount Macedon, Port Phillip', *Tasmanian Journal of Natural Science, Agriculture, Statistics &c* 2: 208–10.
Hobson, E.C., 1846. On the fossil bones at Mount Macedon – Port Phillip, *Tasmanian Journal of Natural Science, Agriculture, Statistics &c* 2: 311.
Hoffman, Antoni, 1989. Mass extinction: The view of a sceptic, *Journal of the Geological Society, London,* 146: 21–35.
Hoffman, P.F. and D.P. Schrag, 2000. Snowball Earth, *Scientific American* 282(1): 50–58.
Holland, Julian, 1992. Thomas Mitchell and the origins of Australian vertebrate palaeontology, in Richard Owen, Thomas Mitchell and Australian science: A commemorative symposium, ed. David Branagan, *Journal of the Proceedings of the Royal Society of New South Wales* 125(3–4): 103–106.
Horton, David, 1991. *Recovering the Tracks: The Story of Australian Archaeology* (Aboriginal Studies Press: Canberra).
Horton, David, ed., 1994. *The Encyclopaedia of Aboriginal Australia,* 2 vols (Australian Institute of Aboriginal and Torres Strait Islander Studies: Canberra).
Horton, David, 2000. *A Pure State of Nature: Sacred Cows, Destructive Myths and the Environment* (Allen & Unwin: Sydney).
Howchin, W., 1895. New facts bearing on the glacial features of Hallett's Cove, *Transactions of the Royal Society of South Australia* 19: 61–70.
Howchin, W., 1901. Preliminary note on the existence of glacial beds of Cambrian age in South Australia, *Transactions of the Royal Society of South Australia* 25: 10.
Howchin, W., 1906. On the geology of the Mt Lofty Ranges, *Transactions of the Royal Society of South Australia* 30: 228.
Howchin, W., 1907. Cambrian and (?) Permo-Carboniferous glaciation in South Australia, *Report of the Eleventh Meeting of the Australasian Association for the Advancement of Science*, Adelaide: 264–272.
Howchin, W., 1908. Glacial beds of Cambrian age in South Australia, *Quarterly Journal of the Geological Society,* London 64: 234–259.
Howchin, W., 1924. Further discoveries of Permo-Carboniferous glacial features near Hallett's Cove, *Transactions of the Royal Society of South Australia* 48: 297–302.
Howitt, A.W., 1904. *The Native Tribes of South-east Australia* (Macmillan and Co: London. Facsimile reprint, Aboriginal Studies Press: Canberra, 1996).
Howitt, A.W. and Otto Siebert, 1902. Two legends of the Lake Eyre tribes, *Report of the Ninth Meeting of the Australasian Association for the Advancement of Science,* Hobart: 525–532.
Hoy, Anthony, 2001. Picking bones, Editor's desk, *Bulletin* 119(32), 25 July 2001.
Hughes, Robin, 1996. Interview with Tom Uren, tape 8, *Australian Biography: Extraordinary Australians Talk About their Lives* (Film Australia 2009). Online <www.australianbiography.gov.au>
Hutton, D. and L. Connors, 1999. *A History of the Australian Environmental Movement* (Cambridge University Press: Cambridge).
Hutton, F.W., 1885. On the supposed glacial epoch in Australia, *Proceedings of the Linnean Society of New South Wales* 10(3): 334–341.
Hutton, James, 1795. *Theory of the Earth with Proofs and Illustrations,* 4 parts (Edinburgh).
Huyssen, Andreas, 1995. *Twilight Memories: Marking Time in a Culture of Amnesia* (Routledge: New York).
Iliffe, J.D. and H. Basedow, 1905. On the formation known as glacial beds of Cambrian age in South Australia, *Transactions of the Royal Society of South Australia* 29: 334.
Iliffe, J.D. and H. Basedow, 1908. On the formation known as 'Glacial beds of Cambrian age' in South Australia, Abstract, *Quarterly Journal of the Geological Society, London* 64: 260.
Imbrie, J. and K. Palmer Imbrie, 1979. *Ice Ages: Solving the Mystery* (Enslow: New Jersey).
Inkster, Ian, 1985. Scientific enterprise and the colonial 'model': Observations on Australian experience in historical context, *Social Studies of Science* 15(4) November: 677–704.

Jameson, Robert, 1831. On the fossil bones found in the bone-caves and bone-breccia of New Holland, *Edinburgh New Philosophical Journal* 10: 393–396 (including a report by Mr William Clift of the College of Surgeons, London, in regard to the fossil bones found in the caves and breccias of New Holland).

Jinman, M., 1994. World Heritage and fossils, *Records of the South Australian Museum* 27, 213–214.

John, Brian, ed., 1979. *The Winters of the World: Earth Under the Ice Ages* (Jacaranda Press: Milton, Queensland).

Johnson, George, 1996. Indian tribes' creationists thwart Archaeologists, *New York Times*, 22 October.

Johnston, Harvey and Peter Clark, 1998. Willandra Lakes archaeological investigations 1968–98, *Archaeology in Oceania* 33(3), Special Issue: 105–119.

Jones, R., 1973. Emerging picture of Pleistocene Australians, *Nature* 246: 278–281.

Jones, Steve, 1999. *Almost Like a Whale: The Origin of Species Updated* (Doubleday: London).

Jose, Nicholas, 1997. *The Custodians* (Pan Macmillan: Sydney).

Jopson, Debra, 1999. Lay ancient Mungo's bones to final rest, say custodians, *Sydney Morning Herald*, 31 May: 5.

Joyce, E.B., 1989. Geological features of significance in Australia. Recent work by the GSA, *The Australian Geologist* 73: 9.

Joyce, E.B., 1994. Geological Heritage Committee, in *Rock Me Hard … Rock Me Soft … A History of the Geological Society of Australia Incorporated*, eds B.J. Cooper and D.F. Branagan (Geological Society of Australia: Sydney), 30–38.

Kennedy, Brian and Barbara Kennedy, 1989. *Australian Place Names* (Hodder and Stoughton: Sydney).

Kohlstedt, Sally Gregory, 1983. Australian museums of natural history: Public priorities and scientific initiatives in the nineteenth century, *Historical Records of Australian Science* 5(4): 1–29.

Kohn, Marek, 1996. *The Race Gallery: The Return of Racial Science* (Vintage: London).

Krefft, Gerard, 1873. Australian natural history: Mammals of Australia and their classification. Part 1. Ornithodelphia and Didelphia, *Transactions of the Royal Society of New South Wales*: 135–148.

Kuhn, Thomas, 1970 (1962). *The Structure of Scientific Revolutions*, 2nd edition (University of Chicago Press: Chicago).

Land Titles Office of South Australia, no date. *Land Services History Book* vol. 35, Adelaide.

Lane, Edward A., 1962. J.T. Woods' Observations on caves, particularly those of South Australia, 1862: Extracts from *Geological Observations in South Australia, Helictite: Journal of Australian Cave Research* 1(1), October: 15–20.

Lang, J.D., 1831. An account of the discovery of bone caves in Wellington Valley, about 210 miles west from Sydney in New Holland, *Edinburgh New Philosophical Journal* 10: 364–368.

Larson, Peter and Kristin Donnan, 2002. *Rex Appeal: The Amazing Story of Sue, the Dinosaur that Changed Science, the Law and My Life* (Invisible Cities Press: Montpelier, Vermont).

Laseron, Charles Francis, 1953. *The Face of Australia* (Angus & Robertson: Sydney).

Laseron, Charles Francis, 1954. *Ancient Australia: The Story of its Past Geography and Life* (Angus & Robertson: Sydney).

Le Grand, H.E., 1988. *Drifting Continents and Shifting Theories: The Modern Revolution in Geology and Scientific Change* (Cambridge University Press: Cambridge).

von Lendenfeld, R., 1885. The glacial period in Australia, *Proceedings of the Linnean Society of New South Wales* 10(1): 44–54.

Levell, David (words) and Michael Wee (photography), 2009. Mungo moondance, *The Australian Way* Qantas in flight magazine, March 2009: 44–52.

Levitt, Norman, 2001. Kennewick Man: burying the truth about America's past, *Spiked Culture* 9 January. Online <www.spiked-online.com/Articles>.

Lloyd, Tim, 2003. Rewriting the book on first Australians, *Advertiser* (Adelaide), 8 March: 35.

Long, John, 2002. *The Dinosaur Dealers* (Allen & Unwin: Sydney).

Long, John and Brian Mackness, 1994. Studies of the Late Cainozoic diprotodontid marsupials of Australia: 4. The Bacchus Marsh Diprotodons, geology, sedimentology and taphonomy, *Records of the South Australian Museum* 27: 95–110.

Lubbock, John, Lord Avebury, 1865. *Prehistoric Times, as Illustrated by Ancient Remains, and the Manners and Customs of Modern Savages* (Williams and Norgate: London).
Ludbrook, N.H., 1980. *A Guide to the Geology and Mineralogy Resources of South Australia* (Government Printer: Adelaide).
Lyell, Charles, 1830. *Principles of Geology, Being an Attempt to Explain the Former Changes of the Earth's Surface by Reference to Causes now in Operation*, vol. 1 (John Murray: London).
Lyell, Charles, 1859. On the occurrence of works of human art in post-Pliocene deposits, *Report of the British Association for the Advancement of Science* 29: 93–95.
Lyell, Charles, 1863. *The Geological Evidences of the Antiquity of Man: With Remarks on Theories of the Origin of Species by Variation* (George W. Childs: Philadelphia).
Lyell, Charles, 1873. *The Geological Evidences of the Antiquity of Man: With an Outline of Glacial and Post-Tertiary Geology and Remarks on Theories of the Origin of Species with Special Reference to Man's First Appearance on the Earth*, 4th edition, revised (John Murray: London).
Lyell, C., 1881. *Life, Letters, and Journals*, ed. K.M. Lyell, 2 vols (John Murray: London).
Lyell, Charles, 1970. *Scientific Journals on the Species Question*, ed. L.G. Wilson (Yale University Press: London).
Maasch, Kirk A., 2003. The big chill, *Cracking the Ice* (WGBH Educational Formation). Online <www.pbs.org/wgbh/nova/ice/chill.html>.
Mabbutt, J.A., 1965. Landscapes of arid and semi-arid Australia, *Australian Natural History* 15: 105–110.
Mabbutt, J.A., 1977. *Desert Landforms: An Introduction to Systematic Geomorphology, Vol. 2* (ANU Press: Canberra).
McBriar, E.M., 2000. The protection of Hallett Cove as a geological monument, in *Records and Reminiscences: Geosciences at the University of Adelaide, 1875–2000* (University of Adelaide: Adelaide), 79–80.
McBryde, Isabel, ed., 1985. *Who Owns the Past?* (Oxford University Press: Melbourne).
McBryde, Isabel, 1996. Past and present indivisible? Archaeology and society, archaeology in society, in *Prehistory to Politics: John Mulvaney, the Humanities and the Public Intellectual*, eds T. Bonyhady and T. Griffiths (Melbourne University Press: Melbourne), 63–84.
MacDonald, Janine, 1999. Mungo Man older than thought, *Age* (Melbourne), 21 May.
McLean, Ian, 1998. *White Aborigines: Identity Politics in Australian Art* (Cambridge University Press: Cambridge).
McLean, Ian, 1998. *Post* colonial: Return to sender, *Australian Humanities Review* December. Online <www.lib.latrobe.edu.au/AHR/archae/Issue-December-1998/mclean2.html>.
MacLeod, Roy, 1987. On visiting the 'moving metropolis': Reflections on the architecture of imperial science, in *Scientific Colonialism, a Cross-cultural Comparison*, eds N. Reingold and M. Rothenberg (Smithsonian Institution Press: Washington DC), 217–251.
McPhee, John, 1998. *Annals of the Former World* (Farrar, Straus and Giroux: New York).
Macumber, P.G., 1969. Interrelationship between physiography, hydrology, sedimentation and salinization of the Loddon River Plains, Australia, *Journal of Hydrology* 7, 39–57.
Manidis Roberts Consultants, 1996. *Sustaining the Willandra: The Willandra Lakes World Heritage Property Plan of Management Summary.* Report to the Australian Government Department of the Environment, Sport and Territories, Canberra.
Manning, Geoffrey H., 1990. *Manning's Place Names of South Australia* (G.H. Manning: Adelaide).
Marsden, Barry M., 1984. *Pioneers of Prehistory: Leaders and Landmarks in English Archaeology (1500–1900)* (G.W. & A. Hesketh: Ormskirk).
Martin, Paul S. and H.E. Wright, Jr, ed., 1967. *Pleistocene Extinctions: The Search for a Cause* (Yale University Press: New Haven).
Martin, Paul S. and Richard G. Klein, eds, 1984. *Quaternary Extinctions: A Prehistoric Revolution* (University of Arizona Press: Tucson).
Mather, P., ed., 1986. *A Time for a Museum: The History of the Queensland Museum, 1862–1986, Memoirs of the Queensland Museum*, vol. 24 (Brisbane: Queensland Museum).

Mawson, D., 1906. On the geology of the Mt Lofty Ranges, *Transactions of the Royal Society of South Australia* 30: 228.
Mawson, D., 1936. Progress in knowledge of the geology of South Australia, *Transactions of the Royal Society of South Australia* 60: lvi–lxv.
Mawson, D., 1938. Walter Howchin, obituary notice, *Proceedings of the Geological Society, London* 94: 118–122.
Mawson, D., 1940. Tillite and other rocks from Hallett Cove, SA, *Transactions of the Royal Society of South Australia* 64: 362.
Meinig, D.W., 1962. *On the Margins of the Good Earth: The South Australian Wheat Frontier, 1869–1884* (Rigby: Chicago).
Miller, Hugh, 1857. *The Testimony of the Rocks, or Geology in its Bearings on the two Theologies, Natural and Revealed* (Shepherd and Elliot: Edinburgh).
Minards, J., 1975. Hallett Cove, *Environment Information Centre Newsletter*, May.
Mincham, Hans, 1967. *Vanished Giants of Australia* (Rigby: Adelaide).
Mitchell, T.L., 1831a. An account of the limestone caves at Wellington Valley, and the situation, near one of them, where fossil bones have been found, *Proceedings of the Geological Society, London* 1(21): 321–324.
Mitchell, T.L., 1831b. Additional information illustrative of the natural history of the Australian bone caves and osseous breccia, *Edinburgh New Philosophical Journal* 10: 368–371.
Mitchell, T.L., 1838. *Three Expeditions into the Interior of Eastern Australia with Descriptions of the Recently Explored Region of Australia Felix and of the Present Colony of New South Wales*, 2 vols (T. & W. Boone: London).
Mitchell, T.L., 1839 (1838). *Three Expeditions into the Interior of Eastern Australia with Descriptions of the Recently Explored Region of Australia Felix and of the Present Colony of New South Wales*, 2 vols, 2nd edition (T. & W. Boone: London. Facsimile reprint, Libraries Board of South Australia: Adelaide, 1965).
Moore, J.R., 1986. Geologists and interpreters of Genesis in the nineteenth century, in *God and Nature: Historical Essays on the Encounter between Christianity and Science*, eds D.C. Lindberg and R.L. Numbers (University of California Press: Berkeley), 322–350.
Morton, Peter, 1996. *After Light: A History of the City of Adelaide and its Council, 1878–1928* (Wakefield Press: Adelaide).
Moyal, A., 1975. Richard Owen and his influence on Australian zoological and palaeontological science, *Historical Records of Australian Science* 2(2): 41–56.
Moyal, Ann, 1986. *'A Bright & Savage Land': Scientists in Colonial Australia* (Collins: Sydney).
Mudrooroo, Nyoongah, 1994. *Aboriginal Mythology: An A–Z Spanning the History of Aboriginal Mythology from the Earliest Legends to the Present Day* (Thorsons: London).
Mulvaney, John, 1990. Bones of contention, *Bulletin*, 9 October: 104–105.
Mulvaney, John, 1991a. Past regained, future lost: The Kow Swamp Pleistocene burials, *Antiquity* 65(246): 12–21.
Mulvaney, John, 1991b. The alpine cultural heritage in perspective, in *Cultural Heritage of the Australian Alps, Proceedings of the 1991 Symposium*, ed. B. Scougall (Australian Alps Liaison Committee: Canberra), 9–18.
Mulvaney, John and Johan Kamminga, 1999. *Prehistory of Australia* (Allen & Unwin: Sydney).
Mulvaney, John, 2005. From Ice Age human fossils to recognition of traditional Aboriginal ownership. *UniNews* University of Melbourne 14(7), 2–16 May <uninews.unimelb.edu.au/news/2304/uninewsarchive.php?volume=14&number=7&publication=un>.
National Parks and Wildlife Service, South Australia (NPWS), no date. Hallett Cove Conservation Park and the Sandison Reserve, brochure (South Australian Government: Adelaide).
NPWS, no date. A key to the self-guided walk in Hallett Cove Conservation Park and the Sandison Reserve, brochure (South Australian Government: Adelaide).
NPWS, 1992. *Naracoorte Caves Conservation Park Management Plan* (South Australian Government, Adelaide).

Nedin, Chris, 1996. Australian dinosaur fossil theft, 15 October. Online <www.cmnh.org/dinoarch/1996Oct/msg00244.html>.

Newman, Keith, 1944a. Riddle of the sands: Erosion solution may be buried there, *Sydney Morning Herald,* 19 December 1944: 5.

Newman, Keith, 1944b. An artist's journey into Australia's 'lost world', *Sydney Morning Herald,* 16 December 1944: 5.

Newman, Keith, 1944c. Drought lands' varied pattern: Falling birthrate of the trees menaces the future, *Sydney Morning Herald,* 18 December 1944: 5.

Newman, Keith, 1944d. Drought losses incalculable: Rain heals only seared earth, *Sydney Morning Herald,* 20 December 1944: 7.

Newman, Keith, 1944e. Country transformed by water: Reward for irrigation, *Sydney Morning Herald,* 20 December 1944: 7.

Newman, Keith and J.M. Leonard, 1944. Photo-essay, Erosion brings desolation to wide areas of dry west, *Sydney Morning Herald,* 20 December 1944: 6.

Newton, Alfred, 1893. Palaeontological discovery in Australia, letter to the editor, *Nature* 47, 27 April: 606.

Orchiston, D. Wayne, R.N. Miller and R.C. Glenie, 1977. A history of nineteenth century investigations at the Lancefield megafaunal site, *The Artefact* 2(3): 105–122.

O'Riordan, T., A. Shadrake and C. Wood, 1989. Interpretation, participation and National Park planning, in *Heritage Interpretation vol. one: The Natural and Built Environment,* papers delivered at the 2nd World Congress on heritage preservation and interpretation (Belhaven Press: London), 179–189.

Our Art Critic, 1945. R. Drysdale's pictures of erosion belt, *Sydney Morning Herald,* 21 November: 9.

Owen, Richard, 1843. On the discovery of the remains of a mastodontoid pachyderm in Australia, *Annals and Magazine of Natural History* 11: 7–12.

Owen, Richard, 1843. Additional evidence proving the Australian pachyderm described in a former number of the 'Annals' to be a *Dinotherium,* with remarks on the nature and affinities of that genus, *Annals and Magazine of Natural History* 11: 329–333.

Owen, Richard, 1844. Description of a fossil molar tooth of a Mastodon discovered by Count Strzelecki in Australia, *Annals and Magazine of Natural History* 14: 268–271.

Owen, Richard, 1845. *Descriptive and Illustrated Catalogue of the Fossil Organic Remains of Mammalia and Aves contained in the Museum of the Royal College of Surgeons of England* (Royal College of Surgeons: London).

Owen, Richard, 1860. *Palaeontology, or a Systematic Study of Extinct Animals and their Geological Relations* (Adam and Charles Black: Edinburgh).

Owen, Richard, 1877. *Researches on the Fossil Remains of the Extinct Mammals of Australia: With a Notice of the Extinct Marsupials of England* (J. Erxleben: London).

Owen, Richard, 1882. Description of portions of a tusk of a proboscidean mammal (*Notelephas australis,* Owen), *Philosophical Transactions of the Royal Society, London* 173: 777–781.

Oxley, John, 1820. *Journals of Two Expeditions into the Interior of New South Wales* (John Murray: London).

Pardoe, Colin, 1995. Riverine, biological and cultural evolution in southeastern Australia, in *Transitions: Pleistocene to Holocene in Australia, Antiquity* 69, Special Number 265: 696–714.

Parker, K. Langloh, 1953 (1896). *Australian Legendary Tales,* ed. and selected by H. Drake-Brockman (Angus & Robertson: Sydney).

Parkin, L.W., ed., 1969, *Handbook of South Australian Geology* (Geological Survey of South Australia: Adelaide).

Pentland, Joseph, 1832. On the fossil bones of Wellington Valley, New Holland or New South Wales, *Edinburgh New Philosophical Journal* 12(24): 120–121.

Pentland, Joseph, 1833. Observations on a collection of fossil bones sent to Baron Cuvier from New Holland, *Edinburgh New Philosophical Journal* 13: 120–121.

Player, Ann, 1992. Julian Tenison Woods, Richard Owen and ancient Australia, in Richard Owen, Thomas Mitchell and Australian science: A commemorative symposium, ed. David Branagan, *Journal of the Proceedings of the Royal Society of New South Wales* 125(3–4): 107–109.

Pledge, N.S., 1994. Fossils of the lake: A history of the Lake Callabonna excavations, *Records of the South Australian Museum* 27: 65–77.

Plimer, Ian, 1997. *A Journey through Stone: The Chillagoe Story* (Reed: Melbourne).

Porter, R., 1977. *The Making of Geology: Earth Science in Britain, 1660–1815* (Cambridge University Press: Cambridge).

Pyne, Stephen J., 1986. *The Ice: A Journey to Antarctica* (University of Iowa Press: Iowa City).

Pyne, Stephen J., 1998. *How the Canyon became Grand: A Short History* (Viking: New York).

Ragless, Margaret E., 1988. *Dust Storms in China Teacups: Ragless Family Heritage to Australia* (Investigator Press: Adelaide).

Ramsay Smith, W., 1930. *The Myths and Legends of the Australian Aborigines* (George G. Harrap & Co).

Reed, A.W., 1965. *Myths and Legends of Australia* (A.H. & A.W. Reed: Sydney).

Reed, A.W., 1977. *Aboriginal Words and Places* (Rigby: Sydney).

Reingold, N. and M. Rothenberg, eds, 1987. Introduction, in *Scientific Colonialism, a Cross-cultural Comparison* (Washington DC: Smithsonian Institution Press).

Rendle, P., R. Major and E.M. McBriar, 2001. Preserving heritage at Hallett Cove, *Heritage Living: Journal of the National Trust of South Australia*, Autumn: 11–12.

Renfrew, Colin and Paul Bahn, 1991. *Archaeology: Theories, Methods, and Practice* (London: Thames and Hudson).

Reynolds, Steve, 1999. Fossil thefts, *MLSSA newsletter* 261, November (Marine Life Society of South Australia). Online <www.mlsaa.asn.au/nletters/november99.htm>.

Rich, P.V., 1985a. *Genyornis newtoni*, a mihirung, in *Kadimakara: Extinct Vertebrates of Australia*, eds P.V. Rich and G.F. van Tets (Pioneer Design Studio, Melbourne), pp. 188–194.

Rich, P.V., 1985b. Priest-geologists and knighted explorers: A short history of the discovery of vertebrate fossils in Australia, in *Kadimakara: Extinct Vertebrates of Australia*, eds P.V. Rich and G.F. van Tets (Pioneer Design Studio: Melbourne), 17–38.

Rich, P.V. (see also P. Vickers-Rich) and G.F. van Tets, eds, 1985. *Kadimakara: Extinct Vertebrates of Australia* (Pioneer Design Studio: Melbourne).

Rich, Thomas H., Patricia Vickers-Rich, Andrew Constantine, Timothy F. Flannery, Lesley Kool and Nicholas van Klaveren, 1997. A tribosphenic mammal from the Mesozoic of Australia, *Science* 278, November: 1438–1442.

Ride, W.D.L. and A.C. Davis, 1997. Origins and setting: Mammal Quaternary palaeontology in the Eastern Highlands of New South Wales, in *Australian Quaternary Vertebrates*, ed. M.L. Augee, *Proceedings of the Linnean Society of New South Wales* 117: 197–221.

Ritson, J. 1922. Professor Howchin, FGS, *Aldersgate Magazine:* 830–832.

Robin, Libby, 1994. Nature conservation as a national concern: the role of the Australian Academy of Science, *Historical Records of Australian Science* 10(1): 1–24.

Robin, Libby, 1998a. Radical ecology and conservation science: an Australian perspective, *Environment and History* 4(2): 191–208.

Robin, Libby, 1998b. *Defending the Little Desert, the Rise of Ecological Consciousness in Australia* (Melbourne University Press: Melbourne).

Robin, Libby, 2007. *How a Continent Created a Nation* (UNSW Press, Sydney).

Rudwick, M.J.S., 1969. The glacial theory, *History of Science* 8: 136–157.

Rudwick, Martin J.S., 1976 (1972). *The Meaning of Fossils: Episodes in the History of Palaeontology*, 2nd edition (Science History Publications: New York).

Rudwick, Martin J.S., 1995 (1992). *Scenes From Deep Time: Early Pictorial Representations of the Prehistoric World*, 2nd edition (University of Chicago Press: Chicago).

Rudwick, Martin, 1996 (1978). Geological travel and theoretical innovation: The role of 'liminal' experience, *Social Studies of Science* 26 (Historic paper): 143–159.

Rudwick, Martin J.S., 1997. *Georges Cuvier, Fossil Bones and Geological Catastrophes: New Translations and Interpretations of the Primary Texts* (University of Chicago Press: Chicago).
Rudwick, Martin J.S., 2005. *Bursting the Limits of Time: The Reconstruction of Geohistory in an Age of Revolution* (University of Chicago Press: Chicago).
Rupke, Nicolaas, 1992. Richard Owen and the Victorian museum movement, in Richard Owen, Thomas Mitchell and Australian science: A commemorative symposium, ed. David Branagan, *Journal of the Proceedings of the Royal Society of New South Wales* 125(3–4): 133–136.
Ryan, Simon, 1996. *The Cartographic Eye: How Explorers Saw Australia* (Cambridge University Press: Cambridge, UK).
Schama, Simon, 1995. *Landscape and Memory* (HarperCollins: London).
Scott, H.J., 1977. *The Development of Landform Studies in Australia* (Bellbird Publishing Services: Sydney).
Scoular, G., 1879. The geology of the Hundred of Munno Para. Part 1. The Newer Tertiary rocks, *Transactions and Proceedings and Report of the Philosophical Society of Adelaide, South Australia* 2: 60–70.
Scoular, G., 1885. Past climatic changes with special reference to the occurrence of a glacial epoch in South Australia, *Transactions of the Royal Society of South Australia* 8: 36–48.
Seddon, G., 1996. Thinking like a geologist: The culture of geology, Mawson Lecture, *Australian Journal of Earth Sciences* 43: 487–495.
Segnit, R.W., 1940. Geology of the Hallett Cove District with special reference to the distribution and age of the Younger Till, *Transactions of the Royal Society of South Australia* 64: 3–45.
Selwyn, A.R.C., 1859. Geological notes of a journey in South Australia from Cape Jervis to Mount Serle, *Parliamentary Papers of South Australia* 20 (Government Printer: Adelaide).
Selwyn, A.R.C. and G.H.F. Ulrich, 1866. *Notes on the Physical Geography, Geology and Mineralogy of Victoria*, La Trobe Collection, State Library of Victoria, Melbourne, Box 239/3, MS6216.
Semonin, Paul, 2000. *American Monster: How the Nation's First Prehistoric Creature Became a Symbol of National Identity* (New York University Press: New York).
Shawcross, Wilfred, 1998. Archaeological excavations at Mungo, *Archaeology in Oceania* 33(3): 183–200.
Sheets-Pyenson, Susan, 1987. Civilizing by nature's example: The development of colonial museums of natural history, 1850–1900, in *Scientific Colonialism, a Cross-cultural Comparison*, eds N. Reingold and M. Rothenberg (Smithsonian Institution Press: Washington DC), 351–378.
Smith, Bernard, 1989 (1985). *European Vision and the South Pacific*, 2nd edition (Oxford University Press: Melbourne).
Smith, Elizabeth, 2001. Lightning Ridge site: The black opal menagerie, *Riversleigh Notes: Australia's Lost Kingdoms special edition* 45: 7–9.
Smith, Geoffrey, 1997. *Russell Drysdale, 1912–81* (National Gallery of Victoria: Melbourne).
Smith, M.A., E. Williams and R.J. Wasson, 1991. The archaeology of the JSN site: Some implications for the dynamics of human occupation in the Strzelecki Desert during the late Pleistocene, *Records of the South Australian Museum* 25(2): 175–192.
Sober, Elliott, 1988. *Reconstructing the Past: Parsimony, Evolution and Inference* (MIT Press: Cambridge MA).
South Australian Government, 1901. Fossil Reserve, Lake Callabonna. Proclamation by His Excellency the Governor in and over the State of South Australia and its Dependencies in the Commonwealth of Australia, *Government Gazette*, 5 December: 1117.
South Australian Science Teachers Association (SASTA), 1974. Hallett Cove today: Tomorrow?, *SASTA Journal*, September: 32–50.
Sprigg, R.C., 1942. The Geology of the Eden–Moana Fault Block, *Transactions of the Royal Society of South Australia* 66: 185–214.
Sprigg, R.C., 1991. Geological summary, in *A Natural History of the Lake Eyre Region: A Visitor's Guide*, eds F.J. Badman, B.K. Arnold and S.L. Bell (NPWS Northern Consultative Committee: Port Augusta), 11–16.

Stafford, Robert A., 1989. *Scientist of Empire: Sir Roderick Murchison, Scientific Exploration and Victorian Imperialism* (Cambridge University Press: Melbourne).

Stephens, C.G. and R.L. Crocker, 1946. Composition and genesis of lunettes, *Transactions of the Royal Society of South Australia* 70: 302–312.

Stirling, E.C., 1894. The recent discovery of fossil remains at Lake Callabonna, South Australia, *Nature* 50: 184–188, 206–211.

Stirling, E.C., 1900. The physical features of Lake Callabonna, *Memoirs of the Royal Society of South Australia* 1(2): 1–15.

Stirling, E.C., 1907. Reconstruction of Diprotodon from the Callabonna Deposits, South Australia, *Nature* 76(1978): 543–544.

Stirling, E.C. and A.H.C. Zietz, 1896a. Preliminary notes on *Genyornis newtoni*, a new genus and species of fossil struthious bird found at Lake Callabonna, *Transactions of the Royal Society of South Australia* 20(1): 171–190.

Stirling, E.C. and A.H.C. Zietz, 1896b. Description of the bones of the leg and foot of *Genyornis newtoni*, a fossil struthious bird from Lake Callabonna, *Transactions of the Royal Society of South Australia* 20(2): 191–211.

Stirling, E.C. and A.H.C. Zietz, 1896c. The newly discovered extinct gigantic bird of South Australia, *Ibis* 7(2): 430, 593.

Stirling, E.C. and A.H.C. Zietz, 1899a. Preliminary notes on *Phascolonus gigas*, Owen [*Phascolomys gigas*, Owen], and its identification with Sceparnedon, *Transactions of the Royal Society of South Australia* 23: 123–135.

Stirling, E.C. and A.H.C. Zietz, 1899b. Fossil remains of Lake Callabonna, Part I: Description of the manus and pes of *Diprotodon australis*, Owen, *Memoirs of the Royal Society of South Australia* 1(1): 1–40.

Stirling, E.C. and A.H.C. Zietz, 1900. Fossil remains of Lake Callabonna, Part II: *Genyornis newtoni*, Stirling and Zietz, a new genus and species of fossil struthious bird, *Memoirs of the Royal Society of South Australia* 1(2): 41–80.

Stirling, E.C. and A.H.C. Zietz, 1905. Fossil remains of Lake Callabonna, Part III: Description of the vertebrae of *Genyornis newtoni*, Stirling and Zietz, *Memoirs of the Royal Society of South Australia* 1(3): 81–110.

Stirling, E.C. and A.H.C. Zietz, 1913a. Fossil remains of Lake Callabonna, Part IV: Description of some further remains of *Genyornis newtoni*, *Memoirs of the Royal Society of South Australia* 1(4): 111–126.

Stirling, E.C. and A.H.C. Zietz, 1913b. Fossil remains of Lake Callabonna, Part V: On the identity of *Phascolomys (Phascolonus) gigas*, Owen, and *Sceparnadon ramsayi*, Owen; With a description of some of its remains, *Memoirs of the Royal Society of South Australia* 1(4): 127–178.

Stirling, W. [Stirling, E.?], 1893. Note on the discovery of *Diprotodon* remains in Australia, *Report of the British Association for the Advancement of Science* 64: 784.

Stirton, R.A., 1954. Digging Down Under, *Pacific Discovery* 7(2): 2–12.

Stirton, R.A., 1963. *Time, Life and Man: The Fossil Record* (John Wiley & Sons: New York).

Stone, Tim, 2004. Robust and gracile, *Australasian Science* 25(2), March: 18–20.

Street, F. Alayne and A.T. Grove, 1979. Global maps of lake level fluctuations since 30,000BP, *Quaternary Research* 12: 83–118.

Sturt, Charles, 1833. *Two Expeditions into the Interior of Southern Australia During the Years 1828, 1829, 1830 and 1831: With Observations on the Soil, Climate, and General Resources of the Colony of New South Wales*, 2 vols (Smith Elder: London).

Sturt, Charles, 1849. *Narrative of an Expedition into Central Australia, Performed under the Authority of Her Majesty's Government During the Years 1844, 5 and 6: Together with a Notice of the Province of South Australia in 1847*, 2 vols and map (T. & W. Boone: London. Facsimile reprint, Libraries Board of South Australia: Adelaide, 1965).

Sturt, Charles, 1984. *Journal of the Central Australian Expedition 1844–45*, ed. Jill Waterhouse (Caliban Books: London).

Talbot, J.L., and R.W. Nesbitt, 1968. *Geological Excursions in the Mount Lofty Ranges and the Fleurieu Peninsula* (Adelaide: Angus and Robertson).

Tate, R.L., 1878. Anniversary address of the president for 1877–8, *Transactions and Proceedings and Report of the Philosophical Society of Adelaide, South Australia* 1: 11–44.

Tate, R.L., 1879. Anniversary address of the president for 1878–9, *Transactions and Proceedings and Report of the Philosophical Society of Adelaide, South Australia* 2: xxxix–lxxv.

Tate, R.L., 1885. Post Miocene climate in South Australia, *Transactions of the Royal Society of South Australia* 8: 49–59.

Tate, R.L., 1887. Glacial phenomena in South Australia, *Report of the First Meeting of the Australasian Association for the Advancement of Science*, Sydney: 231–232.

Tate, R.L., 1893. Century of geological progress, inaugural address of the president, *Report of the Fifth Meeting of the Australasian Association for the Advancement of Science,* Adelaide: 1–69.

Tate, R.L., 1895. Anniversary address of the president for 1894–95, *Transactions of the Royal Society of South Australia* 19: 266–277.

Tate, R.L., W. Howchin and T.W.E. David, 1895. Report to the Research Committee appointed to collect evidence as to glacial action in Australasia: Evidences of glaciation at Hallett's Cove, *Report of the Sixth Meeting of the Australasian Association for the Advancement of Science,* Brisbane: 315–320.

Taylor, Griffith, 1944. Australia's barren spaces, *Sydney Morning Herald,* 16 December: 8.

Tedford, Richard H., 1973. The Diprotodons of Lake Callabonna, *Australian Natural History,* September: 349–354.

Tedford, Richard H., 1985. The Stirton years 1953–1966: A search for Tertiary mammals in Australia, in *Kadimakara: Extinct Vertebrates of Australia,* eds P.V. Rich and G.F. van Tets (Pioneer Design Studio: Melbourne), 39–59.

Tedford, Richard H., 1991. Vertebrate palaeontology in Australia: The American contribution, in *Vertebrate Palaeontology of Australasia,* eds P. Vickers-Rich, J.M. Monaghan, R.F. Baird and T.H. Rich (Pioneer Design Studio: Melbourne), 45–84.

Tedford, Richard H. and R.T. Wells, 1990. Pleistocene deposits and fossil vertebrates from the 'Dead Heart of Australia', *Memoirs of the Queensland Museum* 28(1): 263–284.

Thorne, A.G., 1971. Mungo and Kow Swamp: Morphological variation in Pleistocene Australians, *Mankind* 8: 85–89.

Thorne, A.G., 1977. Separation or reconciliation? Biological clues to the development of Australian society, in *Sunda and Sahul,* eds J. Allen, J. Golson and R. Jones (Academic Press: London).

Thorne, A.G. and P.G. Macumber, 1972. Discoveries of Late Pleistocene man at Kow Swamp, Australia, *Nature* 238: 316–319.

Thorne, Alan, Rainer Grün, Graham Mortimer, Nigel A. Spooner, John J. Simpson, Malcolm McCulloch, Lois Taylor and Darren Curnoe, 1999. Australia's oldest human remains: Age of the Lake Mungo 3 skeleton, *Journal of Human Evolution* 36: 591–612.

Thorp, James, 1965. The nature of the pedological record in the Quaternary, *Soil Science* 99: 1–8.

Tindale, N.B., 1937. Relationship of the extinct Kangaroo Island culture with cultures of Australia, Tasmania and Malaya, *Records of the South Australian Museum* 6: 39–60.

Tindale, N.B., 1957. Culture succession in South-eastern Australia from Late Pleistocene to the present, *Records of the South Australian Museum* 13: 1–49.

Tindale, N.B., 1968. Nomenclature of archaeological cultures and associated implements in Australia, *Records of the South Australian Museum* 15: 616–640.

Tindale, Norman B., 1974. *Aboriginal Tribes of Australia: Their Terrain, Environmental Controls, Distribution, Limits and Proper Names* (ANU Press: Canberra).

[Toteff, S.], 1986. *Hallett Cove Conservation Park Management Plan* (NPWS, Department of Environment and Planning, Adelaide, South Australia).

Trautmann, Thomas R., 1991. The revolution in ethnological time, *Man* (NS)27: 379–397.

Twidale, C.R. and E.M. Campbell, 1993. *Australian Landforms: Structure, Process and Time* (Gleneagle Publishing: Adelaide).

Tylor, Edward B., 1894. On the Tasmanians as representatives of Palaeolithic man, *Journal of the Anthropological Institute of Great Britain and Ireland* 23: 141–152.
Veevers, J.J., 1990. Tectonic–climatic supercycle in the billion-year plate-tectonic eon: Permian Pangean icehouse alternates with Cretaceous dispersed-continents greenhouse, *Sedimentary Geology* 68: 1–16.
Vickers-Rich, P. (see also P.V. Rich) and Neil W. Archbold, 1991. Squatters, priests and professors: A brief history of vertebrate palaeontology in *Terra Australis*, in *Vertebrate Palaeontology of Australasia*, eds P. Vickers-Rich, J.M. Monaghan, R.F. Baird and T.H. Rich (Pioneer Design Studio: Melbourne), 1–44.
Vickers-Rich, Pat (see also P.V. Rich), Tom Rich, Leaellyn Rich and Tim Rich, 1997. *Australia's Ancient Birds: Little Book of Prehistory* (Kangaroo Press: Sydney).
Waite, E.R., 1917. Results of the South Australian Museum expedition to Strzelecki and Cooper Creeks, September and October 1916, *Transactions of the Royal Society of South Australia* 41: 405–475.
Wallace, A.R., 1876. *The Geographical Distribution of Animals, with a Study of the Relations of Living and Extinct Faunas as Elucidating Past Changes of the Earth's Surface*, vol. 1 (Harper and Brothers: New York).
Wallace, A.R., 1892 (1880). *Island Life, or the Phenomena and Causes of Insular Faunas and Floras, Including a Revision and Attempted Solution of the Problem of Geological Climates*, second edition (Macmillan: New York).
Wallis, Lynley A., Ian Moffat, George Trevorrow and Toni Massey, 2008. Locating places for repatriated burial: A case study from Ngarrindjeri *ruwe*, South Australia, *Antiquity* 82(317): 750–760.
Warburton, J. and E.M. McBriar, 1975. The Hallett Cove affair, *Issue: A Journal of Comment*, October, Department of Adult Education, University of Adelaide: 1–12.
Ward, James, 1893. Former state of the country, *South Australian Register*, 21 July: 4.
Ward, Peter D., 1997. *The Call of Distant Mammoths: Why the Ice Age Mammals Disappeared* (Copernicus: New York).
Wells, R.T., 1975. Reconstructing the past: Excavation in fossil caves, *Australian Natural History* 18(6): 208–211.
Wells, R.T. and R.A. Callen, eds, 1986. *The Lake Eyre Basin: Cainozoic sediments, fossil vertebrates and plants, landforms, silcretes and climatic implications*, Australasian Sedimentologists Group field guide series 4 (Geological Society of Australia: Sydney).
Wells, R.T., D.R. Horton and P. Rogers, 1982. *Thylacoleo carnifex* Owen (THYLACOLEONIDAE): Marsupial carnivore?, in *Carnivorous Marsupials*, ed. M. Archer, *Royal Zoological Society of New South Wales*, February: 573–586.
Wells, Roderick T. and Richard H. Tedford, 1995. *Sthenurus* (Macropodidae: Marsupialia) from the Pleistocene of Lake Callabonna, South Australia, *Bulletin of the American Museum of Natural History, New York* 255: 1–111.
Wheaton, C., 1937. An interpreter, *The Australian Christian Commonwealth* 10 December 1937: 1.
White, A., 2001. Where to find fossils in Australia, *Riversleigh Notes: Australia's Lost Kingdoms special edition* 45: 21–30.
White, Hayden, 1987. *The Content of the Form: Narrative Discourse and Historical Representation* (Johns Hopkins University Press: Baltimore).
White, Mary E., 1997. *Listen … Our Land is Crying. Australia's Environment: Problems and Solutions* (Kangaroo Press: Sydney).
White, Osmar, 1973. *The Super-roo of Mungalongaloo,* illustrated by Jeff Hook (Wren: Melbourne).
Wilford, John Noble, 1999. Archaeology and ancestry clash over skeleton, *New York Times* 9 November.
Wilson, C., ed., 1994. *Earth Heritage Conservation* (Geological Society in association with the Open University: London).
Winchester, Simon, 2001. *The Map that Changed the World: A Tale of Rocks, Ruin and Redemption* (Penguin: London).
Woodford, James, 1999. It's a date – our Mungo Man was here 56,000 years ago, *Sydney Morning Herald*, 21 May.

Woods, J.E., 1860. On some Tertiary rocks in the colony of South Australia, *Quarterly Journal of the Geological Society,* London 63: 253–262.
Woods, J.E.T., 1862. *Geological Observations in South Australia, Principally in the District South-east of Adelaide* (Longman, Green, Longman, Roberts and Green: London).
Woods, J.T., 1867. On the glacial period in Australia, *Transactions and Proceedings of the Royal Society of Victoria* 8: 43–47.
Woods, J.T., 1876. History of Australian Tertiary geology, *Transactions of the Royal Society of Tasmania,* 11 July: 76–79.
Woods, J.T., 1883. Physical structure and geology of Australia, *Proceedings of the Linnean Society of New South Wales,* 7: 371–389.
Yelland, E.M., 1983. *Colonists, Copper and Corn in the Colony of South Australia 1850–51, by an Old Colonist,* revised edition (E.M. Yelland: Adelaide).
York, Derek, 1997. *In Search of Lost Time* (Institute of Physics Publication: London).
Young, R.M., 1985. *Darwin's Metaphor: Nature's Place in Victorian Culture* (Cambridge University Press: Cambridge, UK).

Index

www.ingramcontent.com/pod-product-compliance
Ingram Content Group UK Ltd.
Pitfield, Milton Keynes, MK11 3LW, UK
UKHW061956041026
15591UKWH00016B/1027

9 780643 097049